FLORA OF THE
BRITISH ISLES

ILLUSTRATIONS

A. R. CLAPHAM   T. G. TUTIN   E. F. WARBURG

FLORA OF THE BRITISH ISLES

ILLUSTRATIONS

PART I

PTERIDOPHYTA–PAPILIONACEAE

DRAWINGS BY

SYBIL J. ROLES

CAMBRIDGE

AT THE UNIVERSITY PRESS

1957

CAMBRIDGE UNIVERSITY PRESS
Cambridge, New York, Melbourne, Madrid, Cape Town, Singapore,
São Paulo, Delhi, Dubai, Tokyo, Mexico City

Cambridge University Press
The Edinburgh Building, Cambridge CB2 8RU, UK

Published in the United States of America by Cambridge University Press, New York

www.cambridge.org
Information on this title: www.cambridge.org/9780521269629

First published 1957
Re-issued 2010

A catalogue record for this publication is available from the British Library

ISBN 978-0-521-04658-9 Hardback
ISBN 978-0-521-26962-9 Paperback

INTRODUCTION

THESE ILLUSTRATIONS of British plants are intended as a companion to the *Flora of the British Isles* and it is hoped that they will help the user of that book in the recognition of species, though it must be emphasised that identifications should always be checked by reference to the text.

The arrangement of species follows the order of the *Flora*, though not every species described there is illustrated, and a few additional plants not mentioned in the first edition will be found. Species have been omitted on two grounds: first when there are two plants which are closely similar in general appearance, e.g. *Tamarix anglica* and *T. gallica*, a drawing of one is included and the ways in which the other differs may be found in the text; the second, and more frequent, case is that of aliens which have proved to be so uncommon that no fresh specimens have been forthcoming. The latter category includes a few introduced plants, such as *Epimedium alpinum* and *Roemeria hybrida*, which have been included in British Floras at least since the days of *English Botany*, but which appear now to be either extremely rare or extinct in this country. The species represented here which are not described in the *Flora* are either recently recognised natives or introduced plants which have become well established. Much work remains to be done in *Salicornia* before all the British species can be clearly defined. The taxa figured here represent some of the better-known and more widely distributed species.

The nomenclature in general follows that of the *Flora* but a few changes have had to be made as the result of recent investigations, mainly by Mr J. E. Dandy, to whom we are again indebted for advice on this intricate subject. In such cases the name used in the *Flora* is given in parentheses.

The drawings, with very few exceptions, have been made from fresh specimens and are intended primarily to show the general appearance of the plant when alive. Some details have been added which, it is hoped, will aid identification, but no attempt has been made to give large-scale drawings of all parts of the plant. Those who require such details are referred to the series of large illustrations by Miss S. Ross-Craig. The small scale of the present illustrations necessitates the use of a conventional, rather than a strictly naturalistic, representation of hairs and other small parts. It is hoped, however, that they will serve to convey the general appearance of the living plant. The specimens from which the drawings have been made are preserved in the herbarium of the University of Leicester.

ACKNOWLEDGEMENTS

We are greatly indebted to the following for supplying many of the plants illustrated in this Part: W. D. Allen, Miss J. Allison, Miss P. Barrett, P. R. Bell, R. C. L. Burges, Miss M. S. Campbell, Miss N. Churchman, Miss A. P. Conolly, R. W. David, Miss E. W. Davies, J. G. Dony, Miss U. K. Duncan, Rev. E. A. Elliot, E. A. Ennion, Mrs M. R. Gilson, R. A. Graham, A. Hackett, M. K. Hanson, Miss J. E. Hibberd, E. K. Horwood, H. M. Hurst, J. E. Lousley, D. McClintock, E. Milne-Redhead, Miss P. A. Padmore, M. E. D. Poore, T. E. D. Poore, J. E. Raven, Mrs N. Saunders, F. A. Sowter, G. M. Spooner, Mrs F. Le Sueur, F. J. Taylor, Mrs F. J. Taylor, S. M. Walters, Miss McCallum Webster and P. F. Yeo.

A.R.C.
T.G.T.
E.F.W.

PART I

PTERIDOPHYTA–PAPILIONACEAE

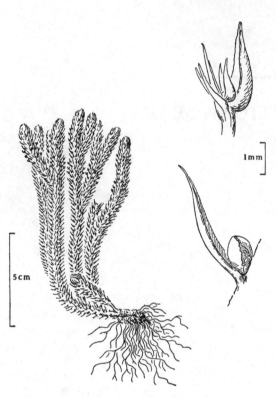

1. *Lycopodium selago* L. 'Fir Clubmoss'

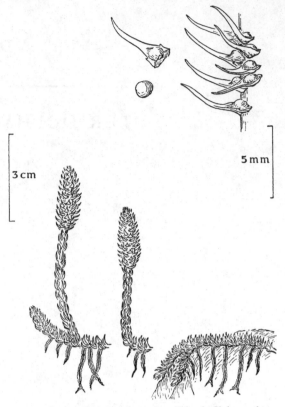

2. *Lycopodium inundatum* L. 'Marsh Clubmoss'

3. *Lycopodium annotinum* L. 'Interrupted Clubmoss'

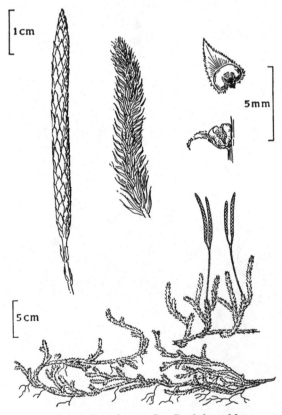

4. *Lycopodium clavatum* L. Stag's-horn Moss

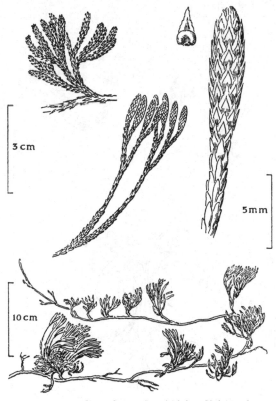

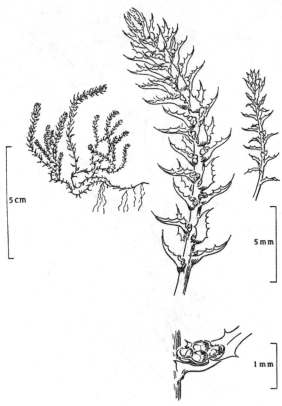

5. *Lycopodium alpinum* L. 'Alpine Clubmoss'

6. *Selaginella selaginoides* (L.) Link 'Lesser Clubmoss'

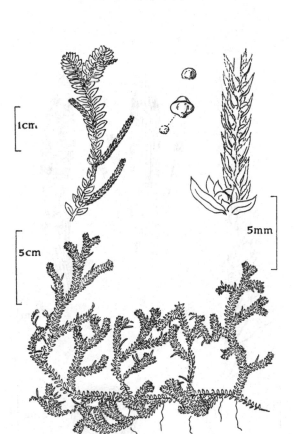

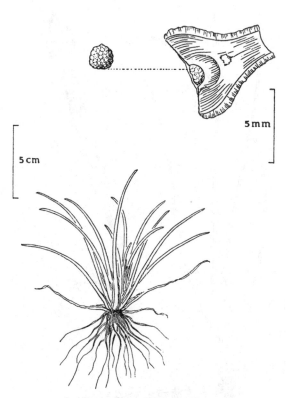

7. *Selaginella kraussiana* (Kunze) A. Braun

8. *Isoetes lacustris* L. Quill-wort

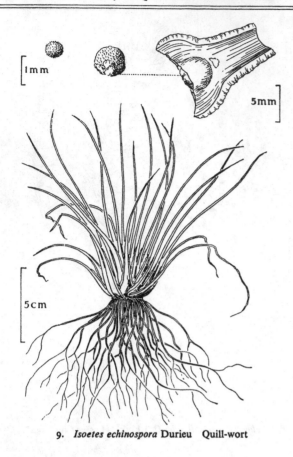

9. *Isoetes echinospora* Durieu Quill-wort

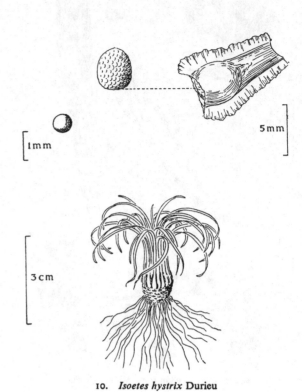

10. *Isoetes hystrix* Durieu

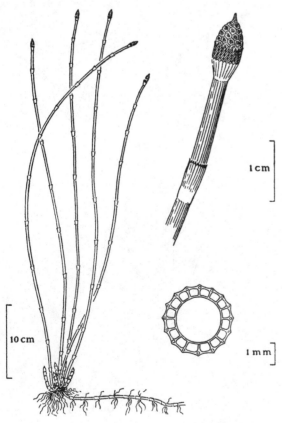

11. *Equisetum hyemale* L. Dutch Rush

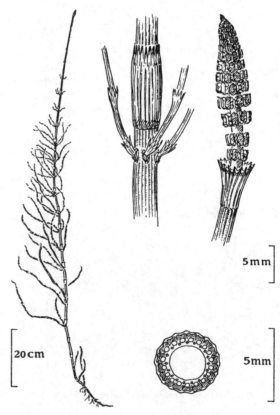

12. *Equisetum ramosissimum* Desf.

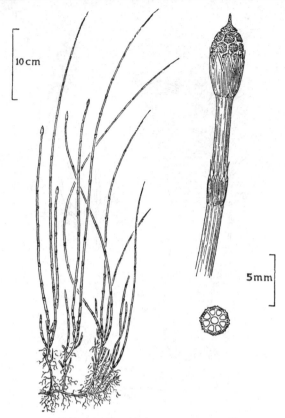

13. *Equisetum × trachyodon* A. Braun   'Mackay's Horsetail'

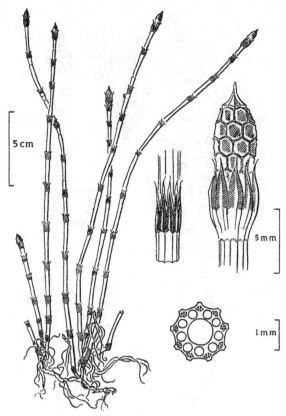

14. *Equisetum variegatum* Schleich.   'Variegated Horsetail'

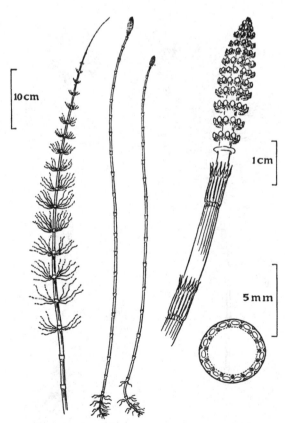

15. *Equisetum fluviatile* L.   'Water Horsetail'

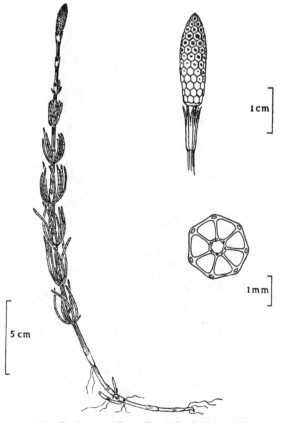

16. *Equisetum palustre* L.   'Marsh Horsetail'

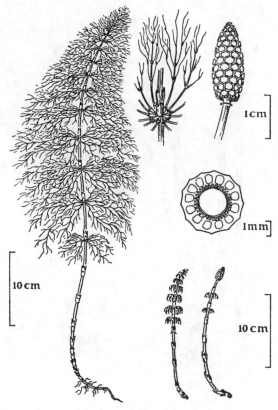

17.   *Equisetum sylvaticum* L. 'Wood Horsetail'

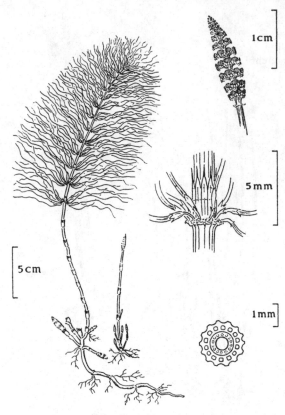

18.   *Equisetum pratense* Ehrh. 'Shady Horsetail'

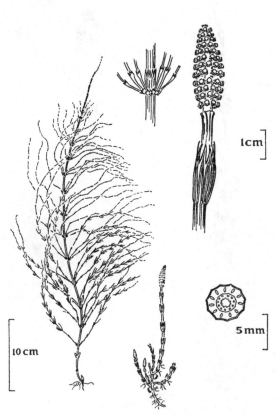

19.   *Equisetum arvense* L. 'Common Horsetail'

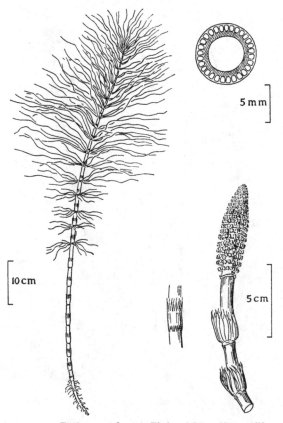

20.   *Equisetum telmateia* Ehrh. 'Great Horsetail'

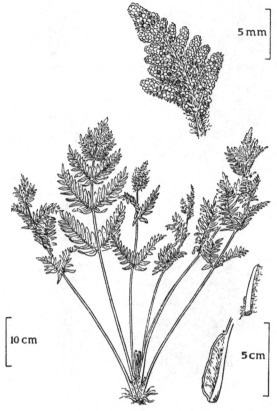

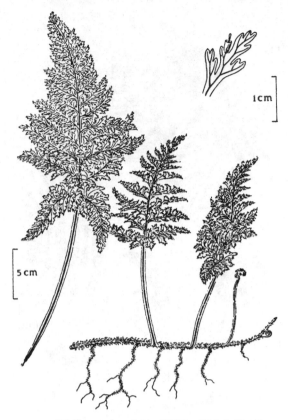

21. *Osmunda regalis* L. Royal Fern

22. *Trichomanes speciosum* Willd. 'Bristle Fern'

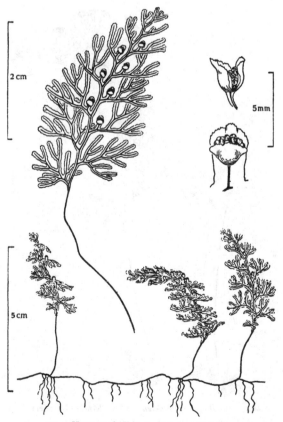

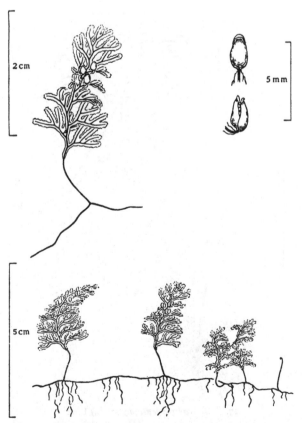

23. *Hymenophyllum tunbrigense* (L.) Sm.
     'Tunbridge Filmy Fern'

24. *Hymenophyllum wilsoni* Hook. 'Wilson's Filmy Fern'

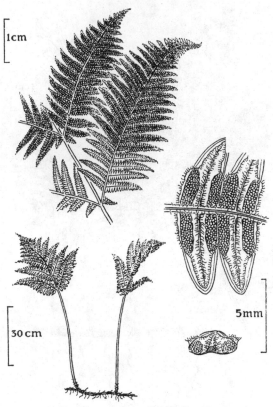

25. *Pteridium aquilinum* (L.) Kuhn   Bracken

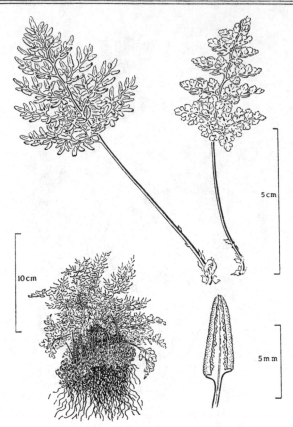

26. *Cryptogramma crispa* (L.) R.Br.   Parsley Fern

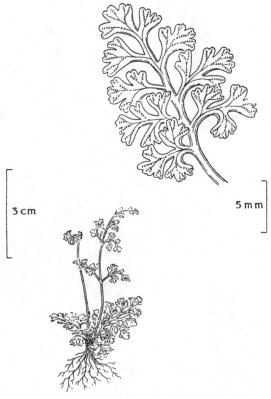

27. *Anogramma leptophylla* (L.) Link

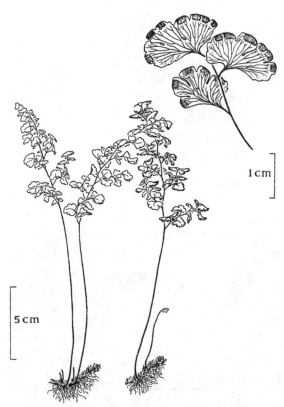

28. *Adiantum capillus-veneris* L.   Maidenhair-fern

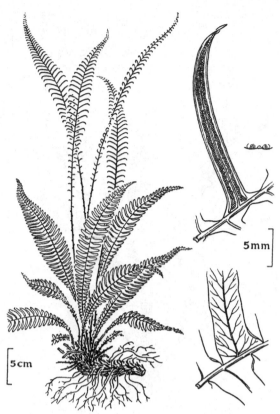

29. *Blechnum spicant* (L.) Roth   Hard-fern

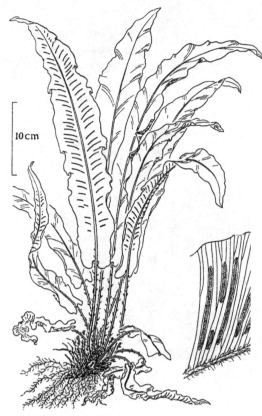

30. *Phyllitis scolopendrium* (L.) Newm.   Hart's-tongue

31. *Asplenium adiantum-nigrum* L.   'Black Spleenwort'

32. *Asplenium onopteris* L.
(*A. adiantum-nigrum* ssp. *onopteris* (L.) Heufl.)

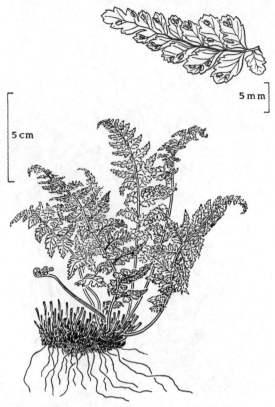

33. *Asplenium obovatum* Viv. 'Lanceolate Spleenwort'

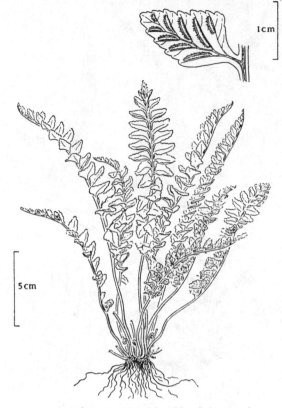

34. *Asplenium marinum* L. 'Sea Spleenwort'

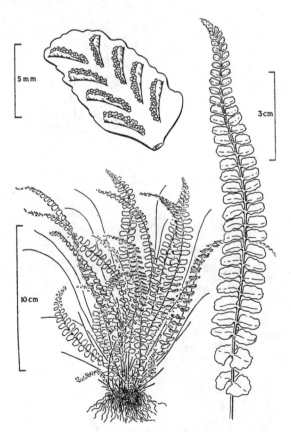

35. *Asplenium trichomanes* L. 'Maidenhair Spleenwort'

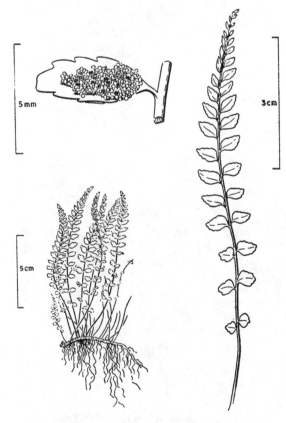

36. *Asplenium viride* Huds. 'Green Spleenwort'

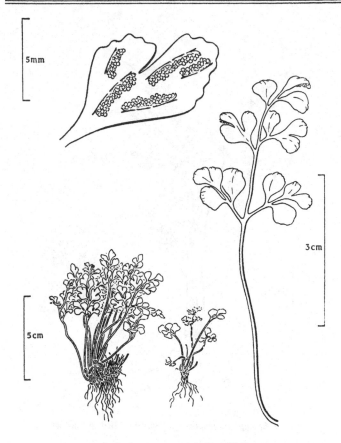

37. *Asplenium ruta-muraria* L.   Wall-rue

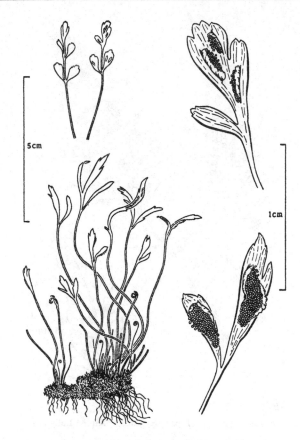

38. *Asplenium* × *alternifolium* Wulf. (above) (*A.* × *breynii*
auct.); *Asplenium septentrionale* (L.) Hoffm.
'Forked Spleenwort' (below)

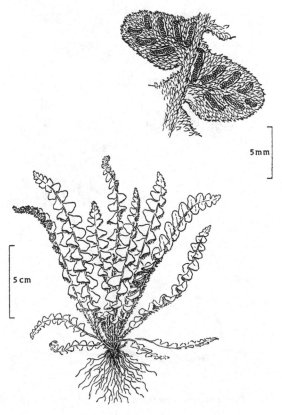

39. *Ceterach officinarum* DC.   Rusty-back Fern

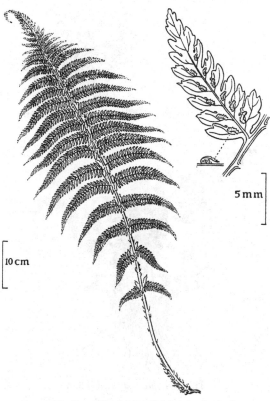

40. *Athyrium filix-femina* (L.) Roth   Lady-fern

3 cm

5 cm

5 mm

41. *Athyrium alpestre* (Hoppe) Rylands   'Alpine Lady-fern'

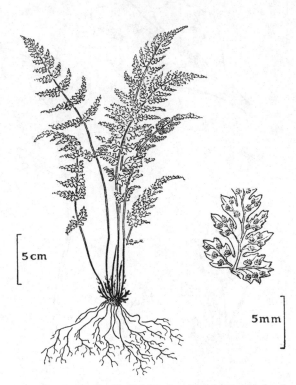

5 cm

5 mm

42. *Cystopteris fragilis* (L.) Bernh.   'Brittle Bladder-fern'

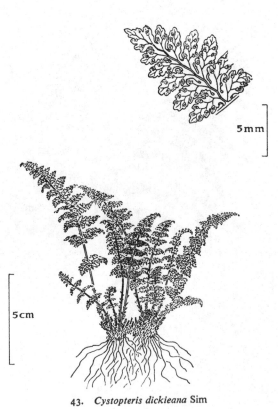

5 mm

5 cm

43. *Cystopteris dickieana* Sim

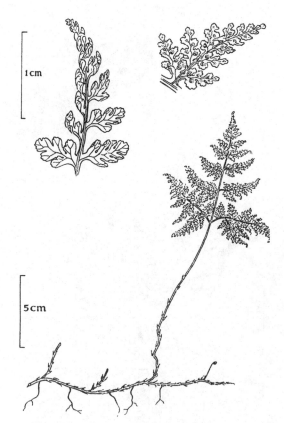

1 cm

5 cm

44. *Cystopteris montana* (Lam.) Desv.   'Mountain Bladder-fern'

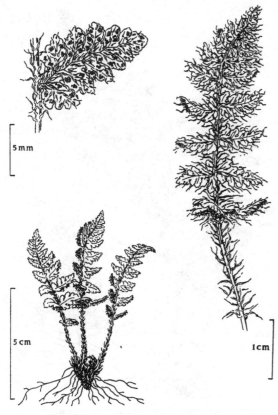

45. *Woodsia ilvensis* (L.) R.Br.

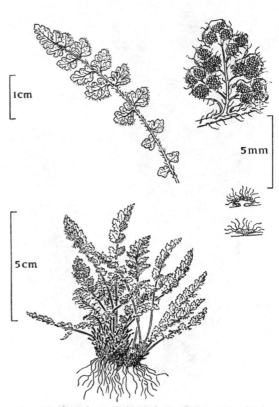

46. *Woodsia alpina* (Bolton) Gray 'Alpine Woodsia'

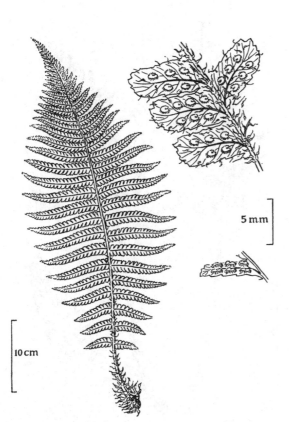

47. *Dryopteris filix-mas* (L.) Schott   Male Fern

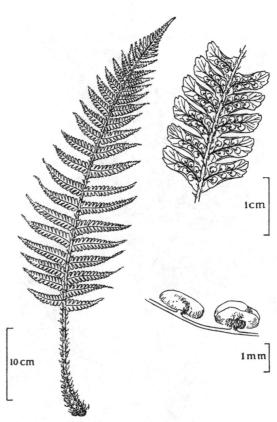

48. *Dryopteris borreri* Newm.

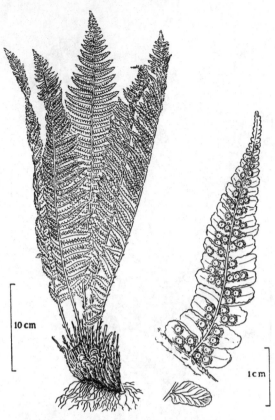

49.   *Dryopteris abbreviata* (DC.) Newm.

50.   *Dryopteris villarsii* (Bell.) Woynar   'Rigid Buckler-fern'

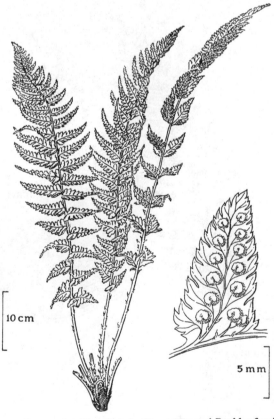

51.   *Dryopteris cristata* (L.) A. Gray   'Crested Buckler-fern'

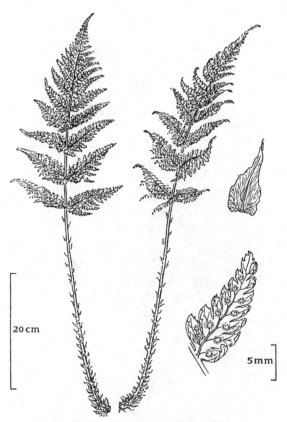

52.   *Dryopteris lanceolatocristata* (Hoffm.) Alst.
(*D. spinulosa* (O. F. Muell.) Watt   'Narrow Buckler-fern'

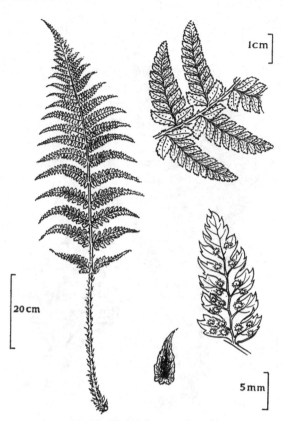

1cm

20cm

5mm

53. *Dryopteris dilatata* (Hoffm.) A. Gray (*D. anstriaca*
(Jacq.) Woynar) 'Broad Buckler-fern'

3mm

5cm

5mm

54. *Dryopteris aemula* (Ait.) Kuntze
'Hay-scented Buckler-fern'

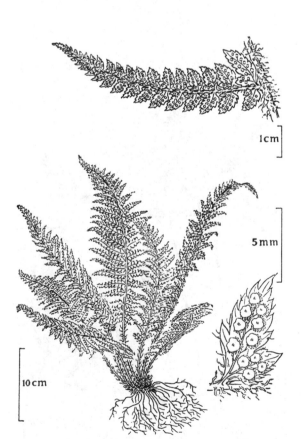

1cm

5mm

10cm

55. *Polystichum setiferum* (Forsk.) Woynar
'Soft Shield-fern'

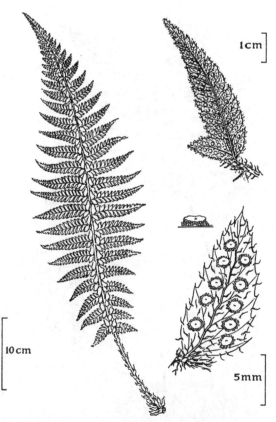

1cm

5mm

10cm

56. *Polystichum aculeatum* (L.) Roth (*P. lobatum* (Huds.)
Chevall.) 'Hard Shield-fern'

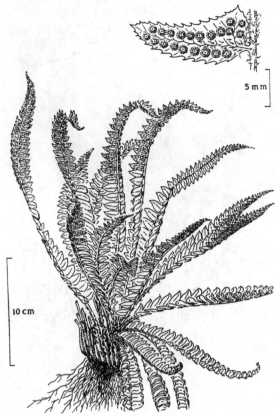

57. *Polystichum lonchitis* (L.) Roth   Holly Fern

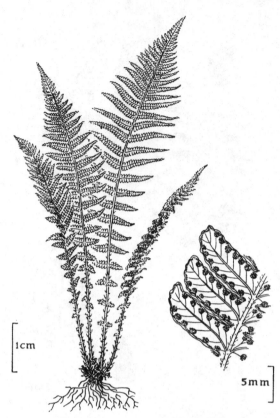

58. *Thelypteris oreopteris* (Ehrh.) Slosson
'Mountain Fern'

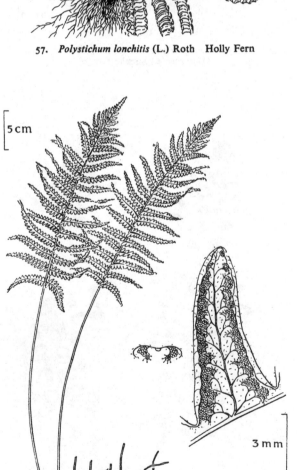

59. *Thelypteris palustris* Schott   'Marsh Fern'

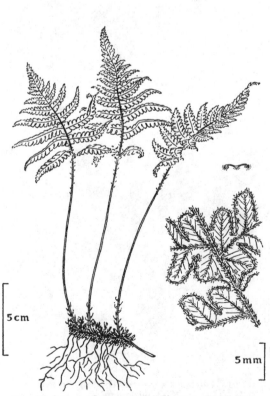

60. *Thelypteris phegopteris* (L.) Slosson   Beech Fern

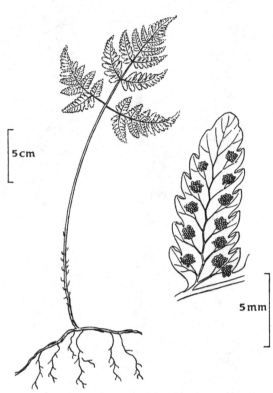

61. *Thelypteris dryopteris* (L.) Slosson   Oak Fern

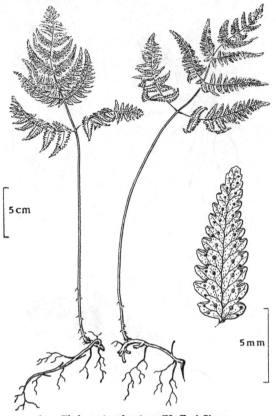

62. *Thelypteris robertiana* (Hoffm.) Slosson
'Limestone Fern'

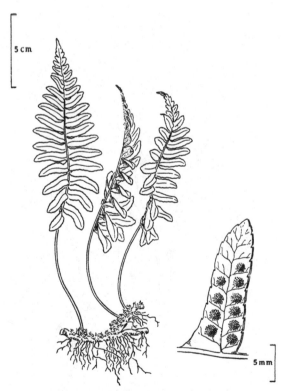

63. *Polypodium vulgare* L. 'Polypody'

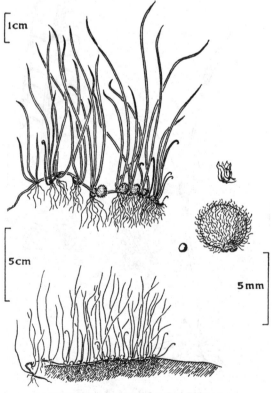

64. *Pilularia globulifera* L.   Pillwort

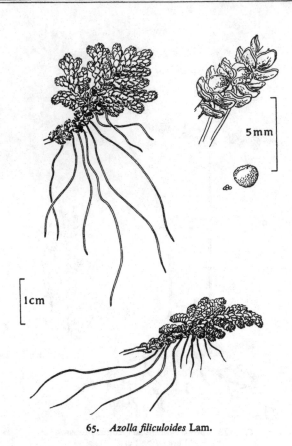

65. *Azolla filiculoides* Lam.

66. *Botrychium lunaria* (L.) Sw.   Moonwort

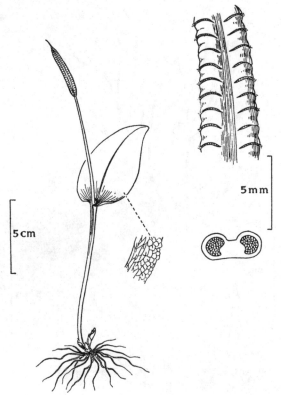

67. *Ophioglossum vulgatum* L. ssp. *vulgatum*
Adder's Tongue

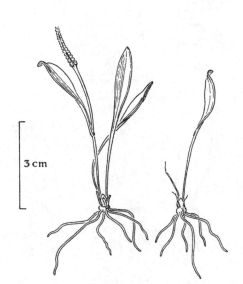

68. *Ophioglossum lusitanicum* L.

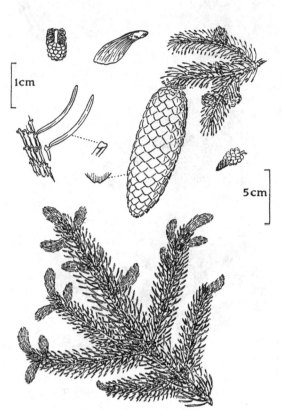

69. *Picea abies* (L.) Karst.   Norway Spruce

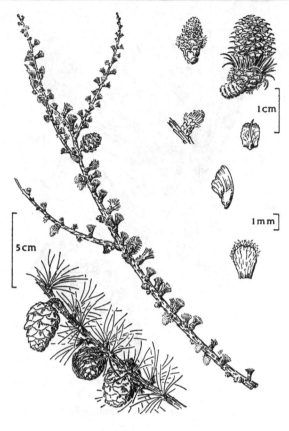

70. *Larix decidua* Mill.   European Larch

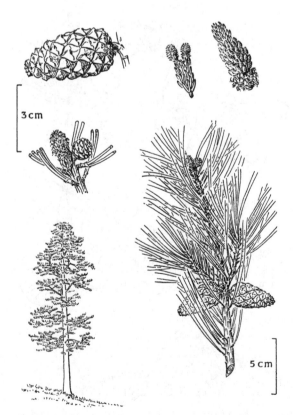

71. *Pinus sylvestris* L.   Scots Pine

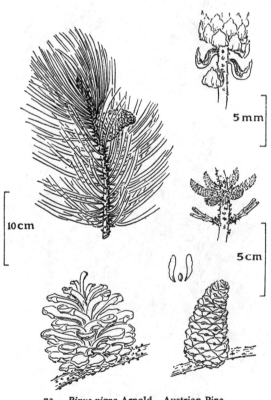

72. *Pinus nigra* Arnold   Austrian Pine

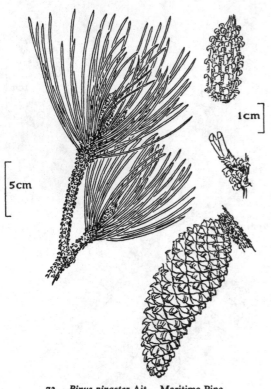

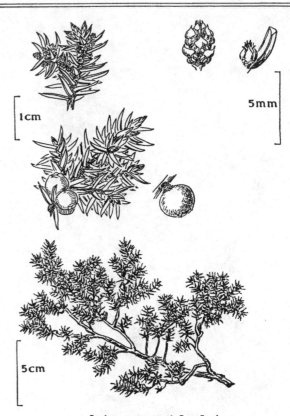

73. *Pinus pinaster* Ait.   Maritime Pine

74. *Juniperus communis* L.   Juniper

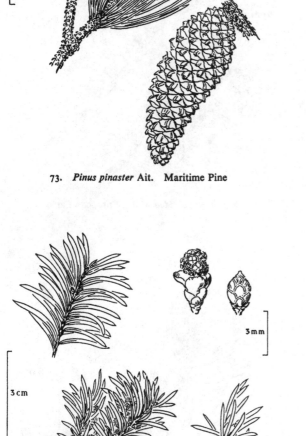

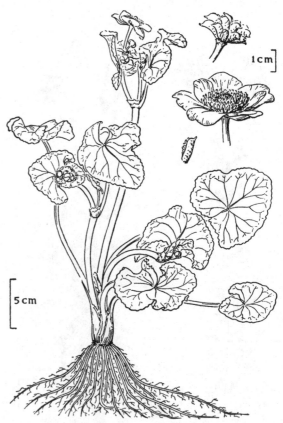

75. *Taxus baccata* L.   Yew

76. *Caltha palustris* L.   Kingcup, Marsh Marigold   Yellow

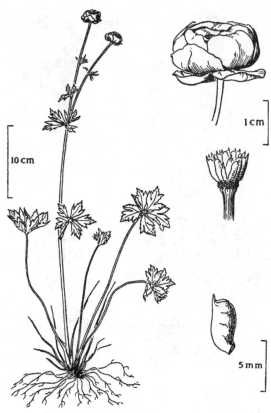

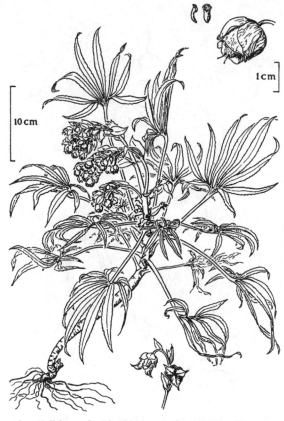

77. *Trollius europaeus* L. 'Globe Flower' Yellow

78. *Helleborus foetidus* L. Bear's-foot Yellowish green

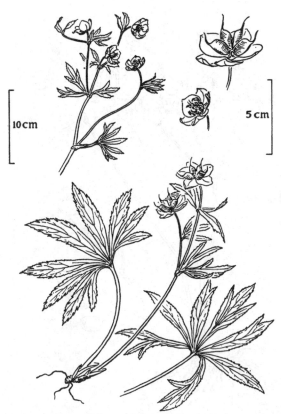

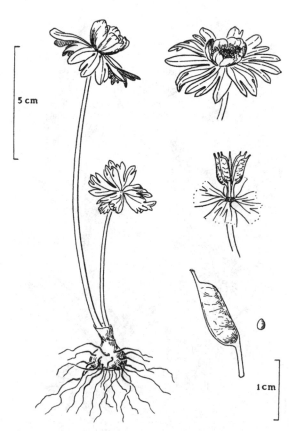

79. *Helleborus viridis* L. Bear's-foot Yellowish green

80. *Eranthis hyemalis* (L.) Salisb.
'Winter Aconite' Yellow

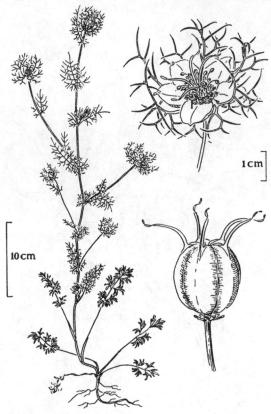

81. *Nigella damascena* L.   Love-in-a-Mist   Blue

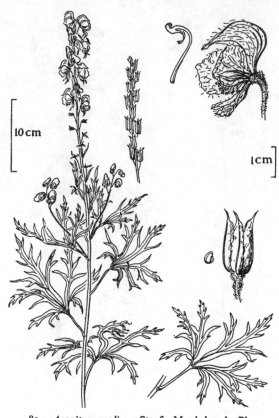

82. *Aconitum anglicum* Stapf   Monkshood   Blue

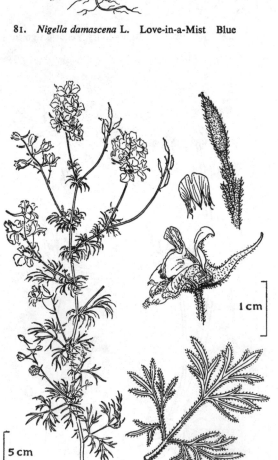

83. *Delphinium gayanum* Wilmott
(*D. ajacis* L. sec. J. Gay)   Larkspur   Blue

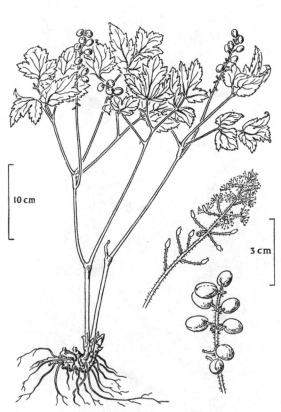

84. *Actaea spicata* L.   Baneberry,
Herb Christopher   White

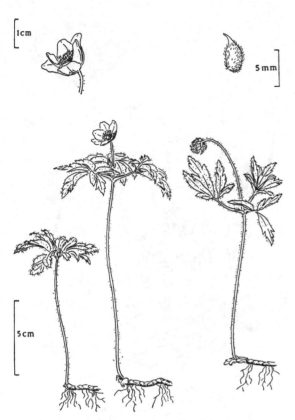

85.   *Anemone nemorosa* L.   'Wood Anemone'
White to purplish

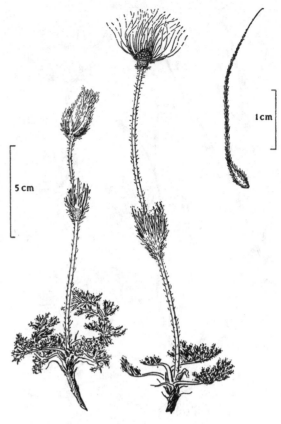

86.   *Pulsatilla vulgaris* Mill.
(*Anemone pulsatilla* L.)   'Pasque Flower'   Purple

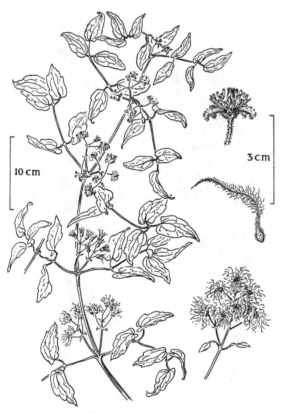

87.   *Clematis vitalba* L.
Traveller's Joy, Old Man's Beard   Greenish white

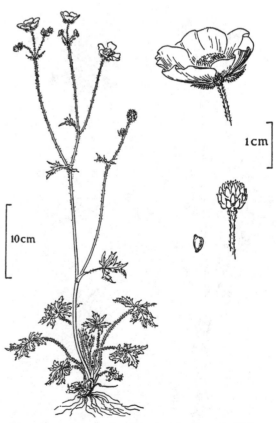

88.   *Ranunculus acris* L.   'Meadow Buttercup'   Yellow

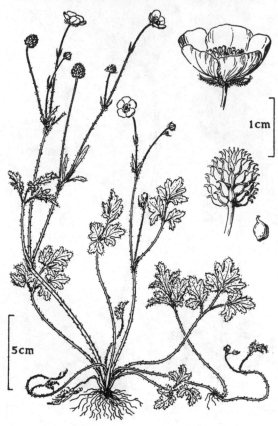

89. *Ranunculus repens* L. 'Creeping Buttercup' Yellow

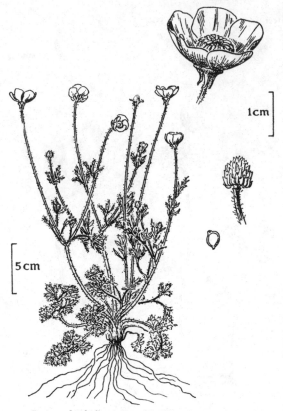

90. *Ranunculus bulbosus* L. 'Bulbous Buttercup' Yellow

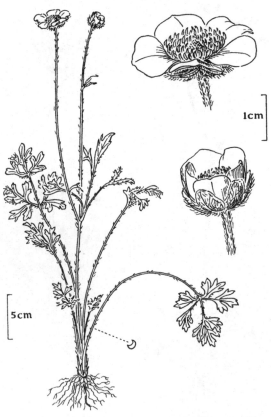

91. *Ranunculus paludosus* Poir. (*R. flabellatus* Desf.)
     'Fan-leaved Buttercup' Yellow

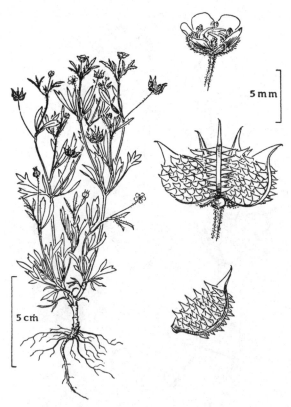

92. *Ranunculus arvensis* L. 'Corn Crowfoot' Yellow

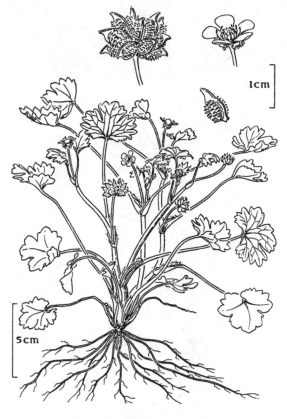

93. *Ranunculus muricatus* L. Yellow

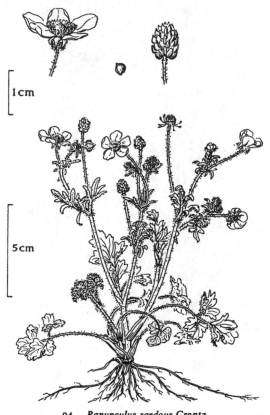

94. *Ranunculus sardous* Crantz
'Hairy Buttercup' Yellow

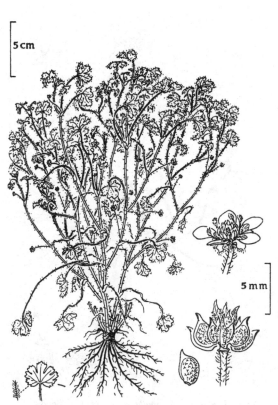

95. *Ranunculus parviflorus* L.
'Small-flowered Buttercup' Yellow

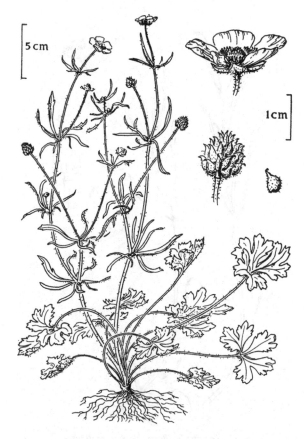

96. *Ranunculus auricomus* L. Goldilocks Yellow

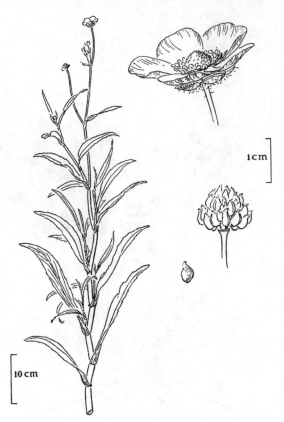

97. *Ranunculus lingua* L.   Great Spearwort   Yellow

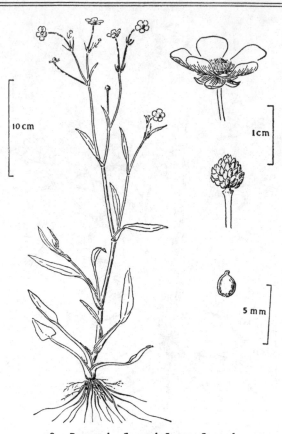

98. *Ranunculus flammula* L. ssp. *flammula*
Lesser Spearwort   Yellow

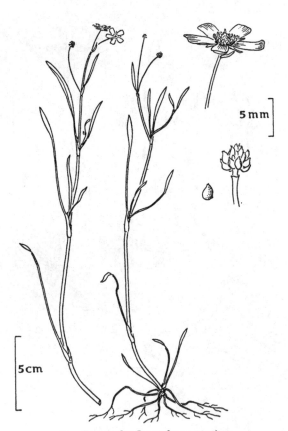

99. *Ranunculus flammula* ssp. *scoticus*
(E. S. Marshall) Clapham   Yellow

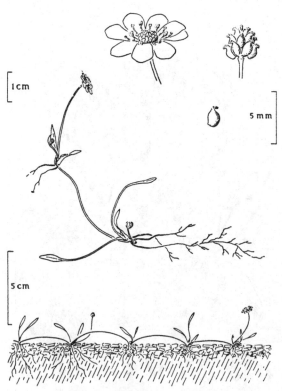

100. *Ranunculus flammula* × *reptans* L.   Yellow

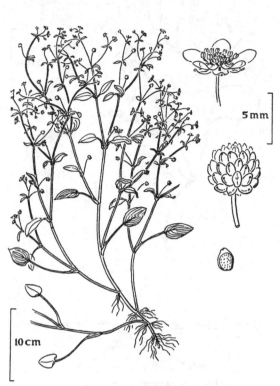

101.   *Ranunculus ophioglossifolius* Vill.
'Snaketongue Crowfoot'   Yellow

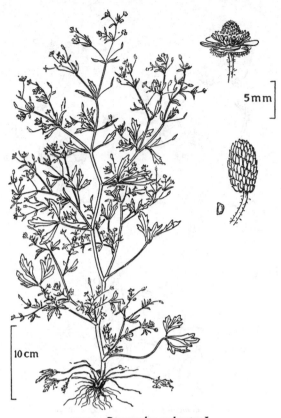

102.   *Ranunculus sceleratus* L.
'Celery-leaved Crowfoot'   Yellow

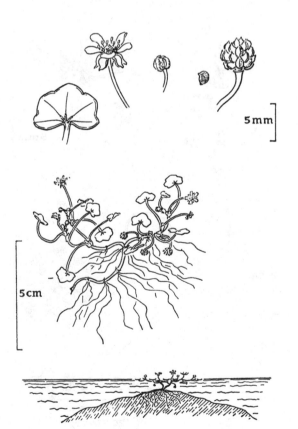

103.   *Ranunculus hederaceus* L.
'Ivy-leaved Water Crowfoot'   White

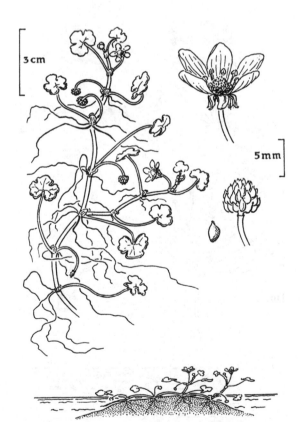

104.   *Ranunculus lenormandi* F. Schultz
'Lenormand's Water Crowfoot'   White

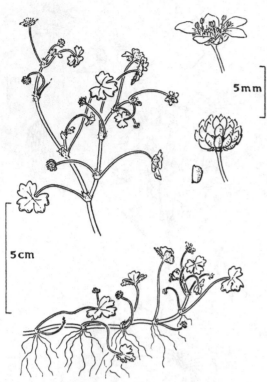

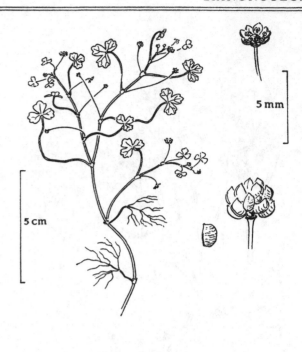

105.   *Ranunculus lutarius* (Revel) Bouvet
'Mud Crowfoot'   White

106.   *Ranunculus tripartitus* DC.
'Three-lobed Water Crowfoot'   White

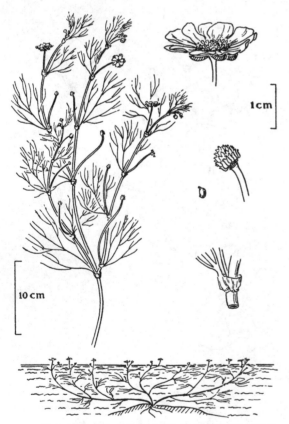

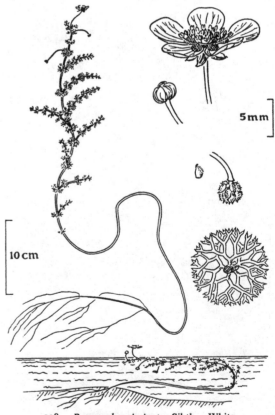

107.   *Ranunculus fluitans* Lam.   'Water Crowfoot'   White

108.   *Ranunculus circinatus* Sibth.   White

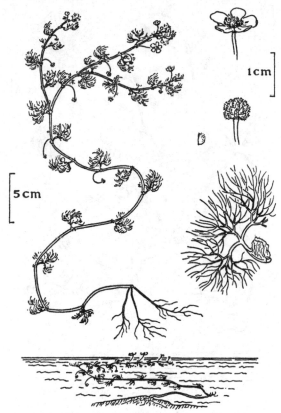

109. *Ranunculus trichophyllus* Chaix ssp. *trichophyllus* White

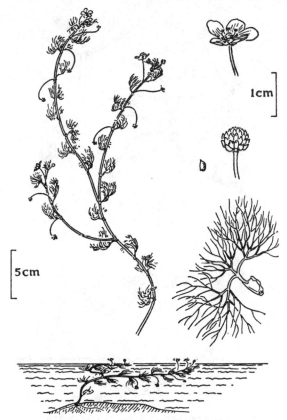

110. *Ranunculus trichophyllus* ssp. *drouetii*
(Godr.) Clapham White

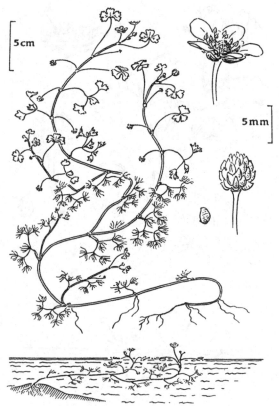

111. *Ranunculus aquatilis* L. ssp. *aquatilis*
'Water Crowfoot' White

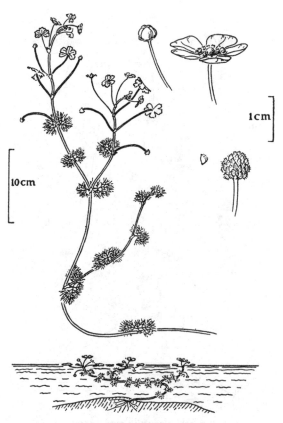

112. *Ranunculus aquatilis* ssp. *radians*
(Revel) Clapham White

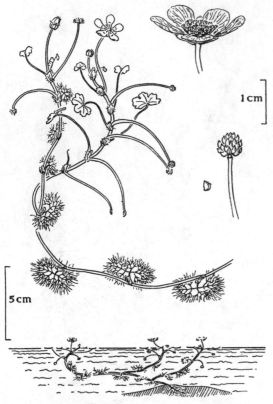

113. *Ranunculus aquatilis* ssp. *peltatus* (Schrank)
Syme   White

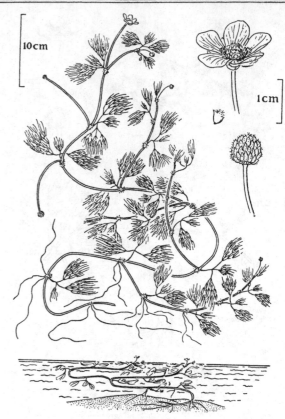

114. *Ranunculus aquatilis* ssp. *pseudofluitans*
(Syme) Clapham   White

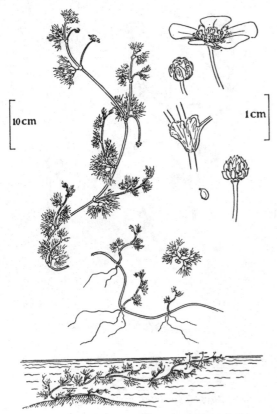

115. *Ranunculus aquatilis* ssp. *sphaerospermus*
Boiss. & Blanche) Clapham   White

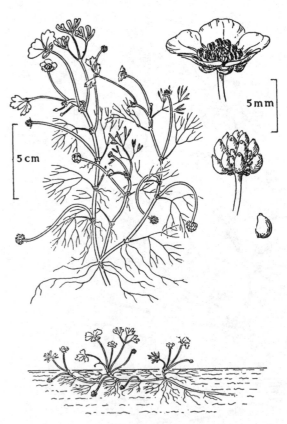

116. *Ranunculus baudotii* Godr.   White

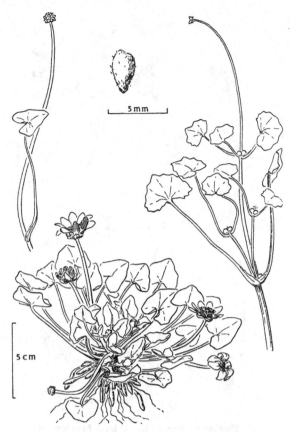

117. *Ranunculus ficaria* L. var. *fertilis* Clapham (left);
var. *ficaria* (right) Lesser Celandine Yellow

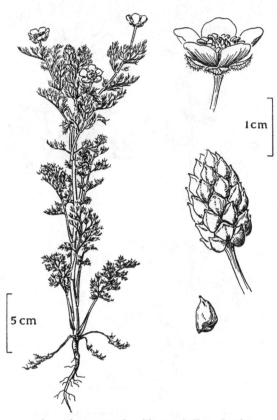

118. *Adonis annua* L. Pheasant's Eye Scarlet

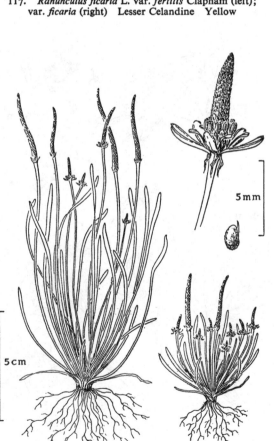

119. *Myosurus minimus* L. Mouse-tail Greenish

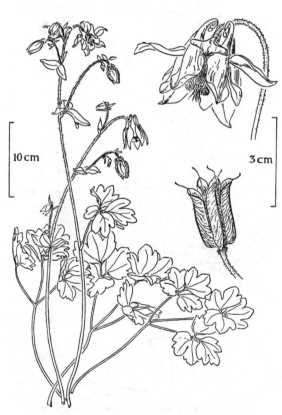

120. *Aquilegia vulgaris* L. Columbine
Blue (white or reddish)

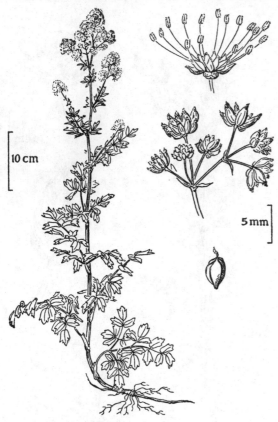

121.   *Thalictrum flavum* L.   'Common Meadow Rue'
Whitish

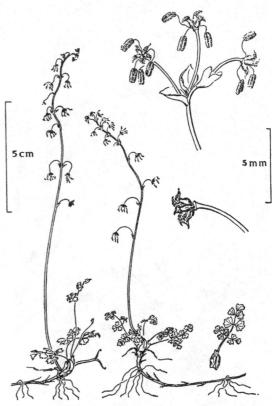

122.   *Thalictrum alpinum* L.   'Alpine Meadow Rue'
Purplish

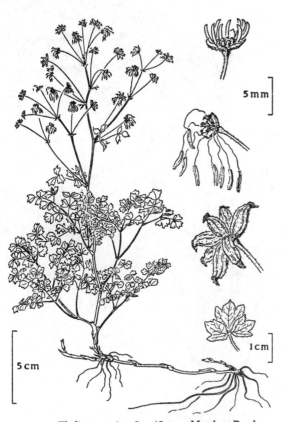

123.   *Thalictrum minus* L.   'Lesser Meadow Rue'
Yellowish or purplish green

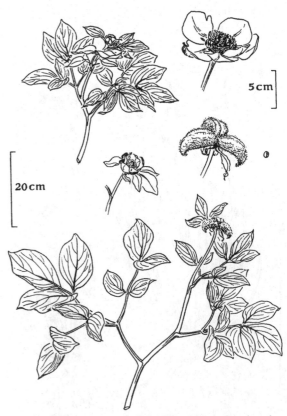

124.   *Paeonia mascula* (L.) Mill.   Peony   Purple-red

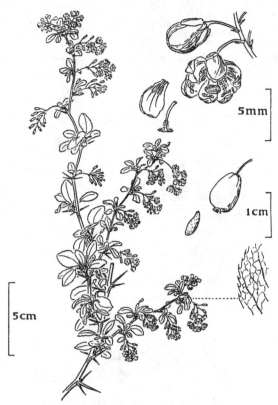

125. *Berberis vulgaris* L.   Barberry   Yellow

126. *Mahonia aquifolium* (Pursh) Nutt.
Oregon Grape   Yellow

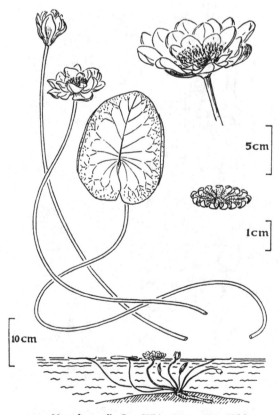

127. *Nymphaea alba* L.   White Water-lily   White

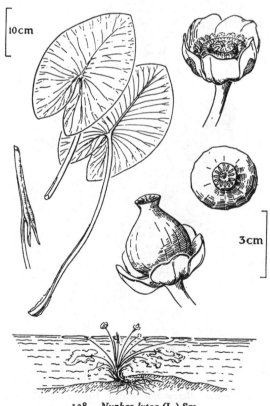

128. *Nuphar lutea* (L.) Sm.
Yellow Water-lily, Brandy-bottle   Yellow

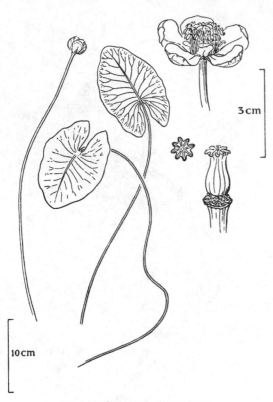

129. *Nuphar pumila* (Timm) DC.
'Least Yellow Water-lily' Yellow

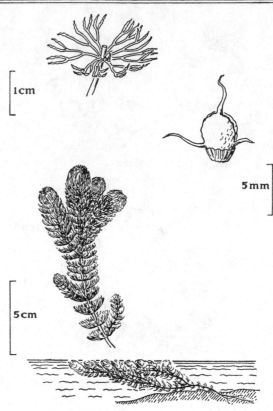

130. *Ceratophyllum demersum* L. Horn-wort Green

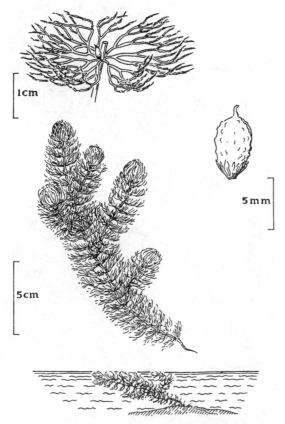

131. *Ceratophyllum submersum* L. Horn-wort Green

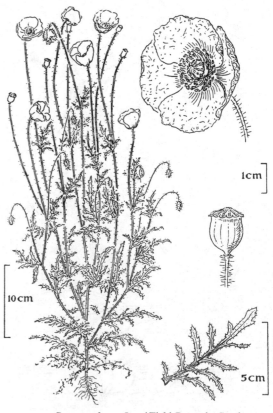

132. *Papaver rhoeas* L. 'Field Poppy' Scarlet

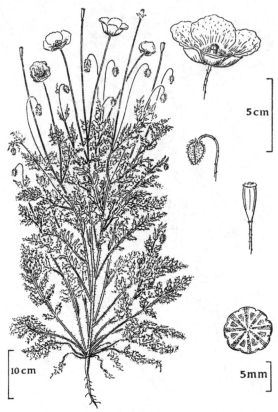

5cm

10cm   5mm

133.   *Papaver dubium* L.   'Long-head Poppy'   Scarlet

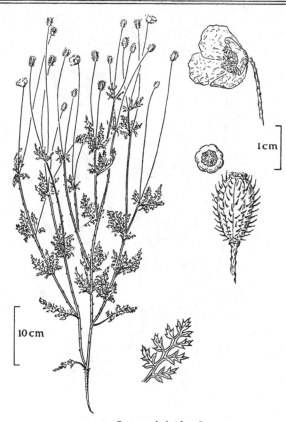

1cm

10cm

134.   *Papaver hybridum* L.
'Round Prickly-headed Poppy'   Crimson

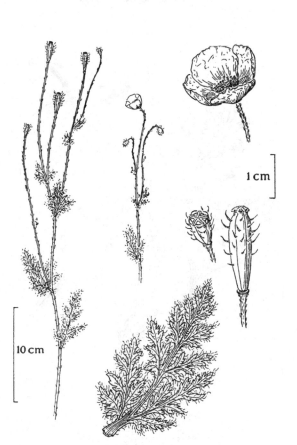

1cm

10cm

135.   *Papaver argemone* L.
'Long Prickly-headed Poppy'   Scarlet

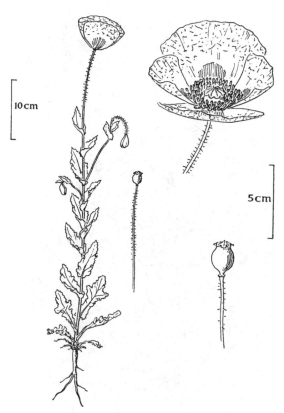

10cm

5cm

136.   *Papaver somniferum* L.   Opium Poppy
White or pale lilac

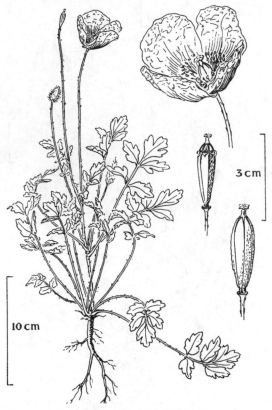

137. *Meconopsis cambrica* (L.) Vig.
Welsh Poppy   Yellow

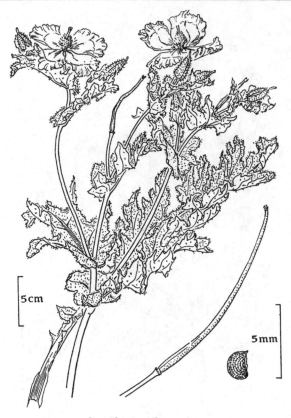

138. *Glaucium flavum* Crantz
'Yellow Horned-poppy'   Yellow

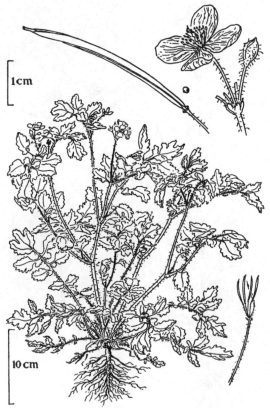

139. *Chelidonium majus* L.   Greater Celandine   Yellow

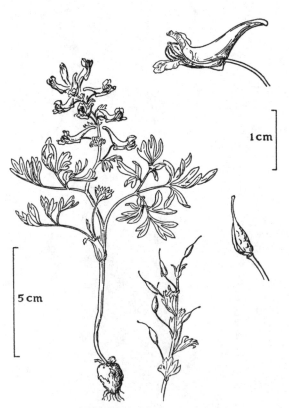

140. *Corydalis solida* (L.) Sw.   Purple

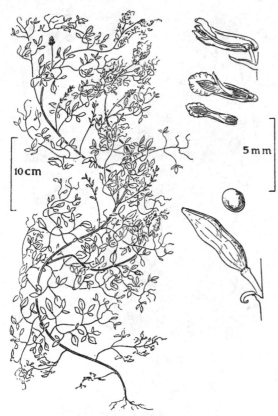

141. *Corydalis claviculata* (L.) DC.
'White Climbing Fumitory' Cream

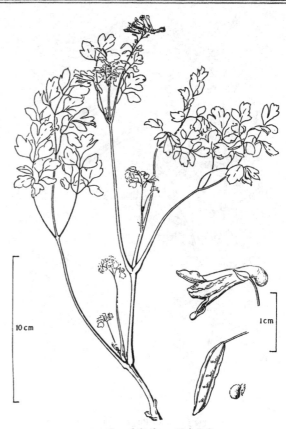

142. *Corydalis lutea* (L.) DC.
'Yellow Fumitory' Yellow

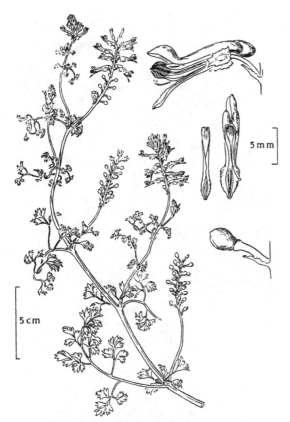

143. *Fumaria occidentalis* Pugsl. White and red

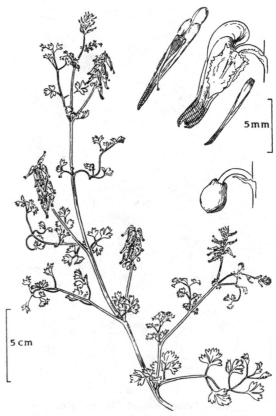

144. *Fumaria capreolata* L. 'Ramping Fumitory'
Cream and dark red

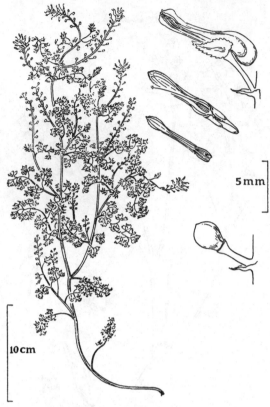

145. *Fumaria purpurea* Pugsl.   Pink and purple

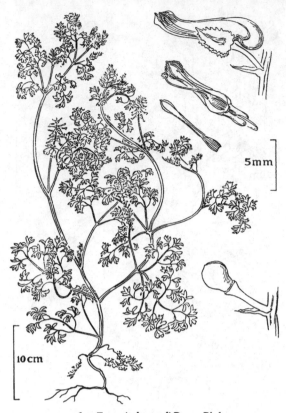

146. *Fumaria bastardi* Bor.   Pink

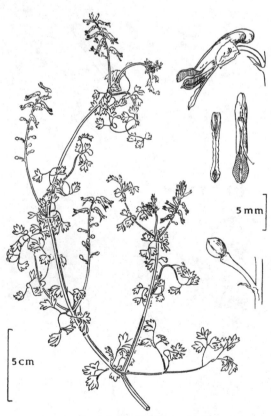

147. *Fumaria martinii* Clavaud   Pink

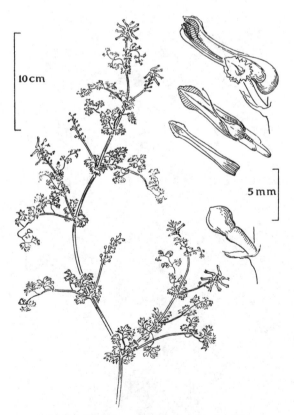

148. *Fumaria muralis* Sonder ssp. *boraei*
     (Jord.) Pugsl.   Pink

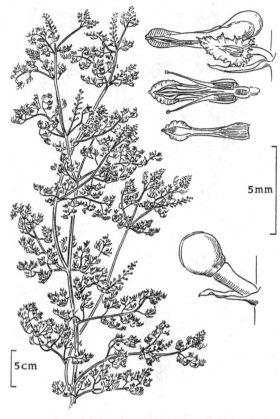

149. *Fumaria micrantha* Lag.   Pink

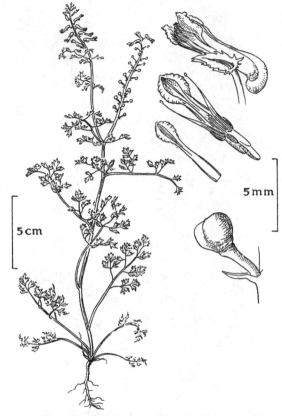

150. *Fumaria officinalis* L.   'Common Fumitory'   Pink

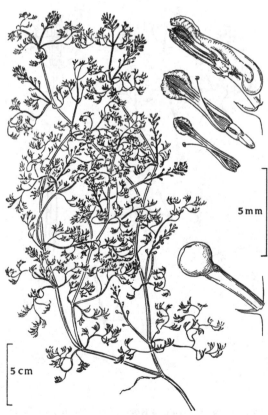

151. *Fumaria vaillantii* Lois.   Pale pink

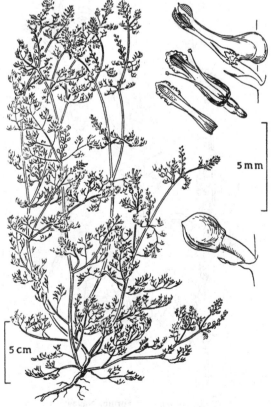

152. *Fumaria parviflora* Lam.   White or pinkish

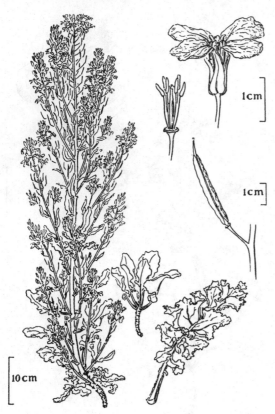

1cm

1cm

153. *Brassica oleracea* L. Wild Cabbage Yellow

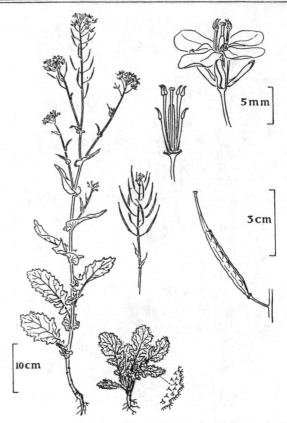

5mm

3cm

154. *Brassica napus* L. Rape, Cole, Swede Yellow or buff

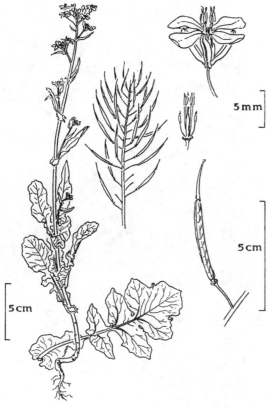

5mm

5cm

155. *Brassica rapa* L. Turnip, Navew Yellow

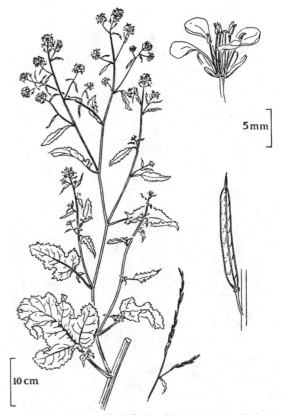

5mm

156. *Brassica nigra* (L.) Koch Black Mustard Yellow

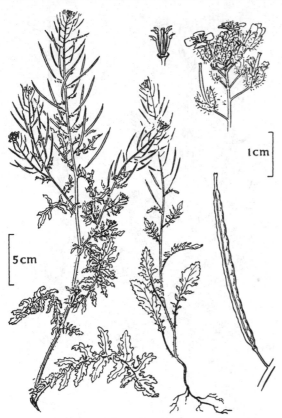

157. *Erucastrum gallicum* (Willd.) O. E. Schulz
Pale yellow

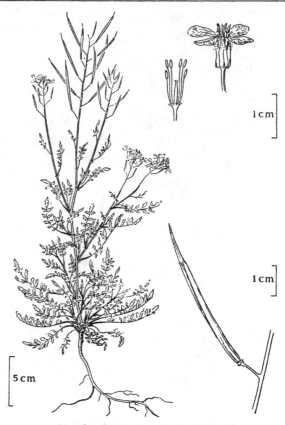

158. *Rhynchosinapis monensis* (L.) Dandy
'Isle of Man Cabbage'   Pale yellow

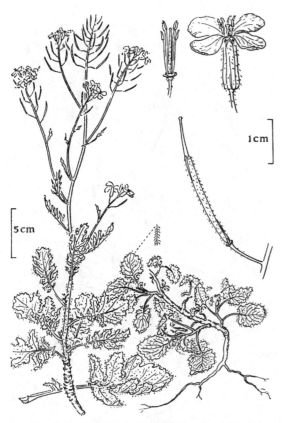

159. *Rhynchosinapis wrightii* (O. E. Schulz) Dandy
'Lundy Cabbage'   Yellow

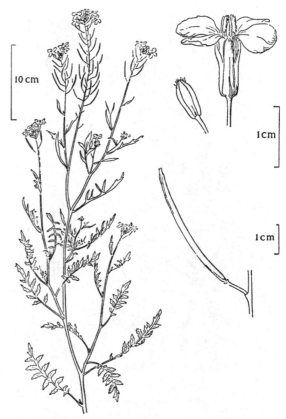

160. *Rhynchosinapis cheiranthos* (Vill.) Dandy
(*R. erucastrum* (L.) Dandy, p.p.)   'Tall Wallflower Cabbage'
Pale yellow

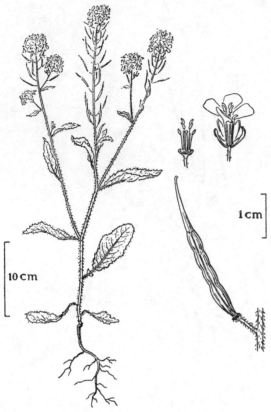

161. *Sinapis arvensis* L. Charlock Yellow

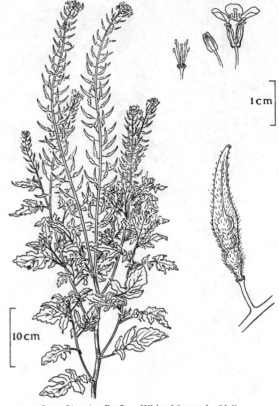

162. *Sinapis alba* L. White Mustard Yellow

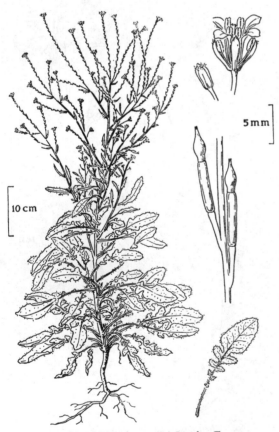

163. *Hirschfeldia incana* (L.) Lagrèze-Fossat
'Hoary Mustard' Pale yellow

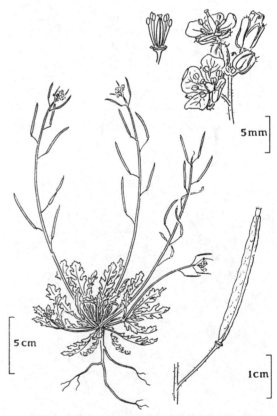

164. *Diplotaxis muralis* (L.) DC. Wall Rocket Yellow

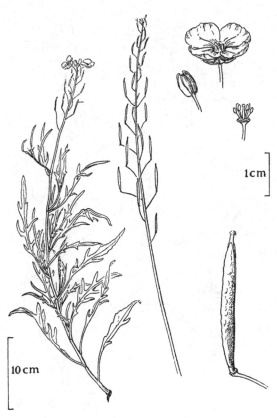

165.　*Diplotaxis tenuifolia* (L.) DC.
'Perennial Wall Rocket'　Yellow

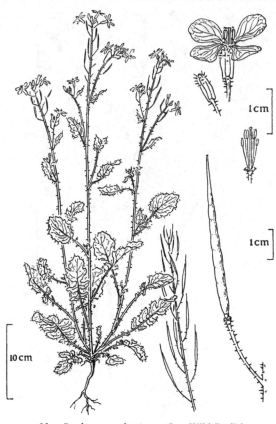

166.　*Raphanus raphanistrum* L.　Wild Radish
Yellow, lilac or white

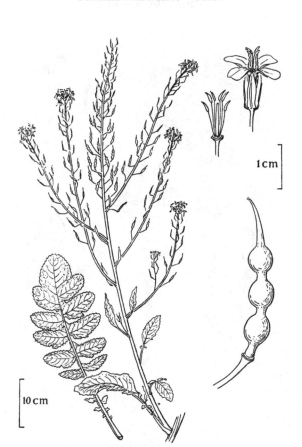

167.　*Raphanus maritimus* Sm.　'Sea Radish'
Yellow or white

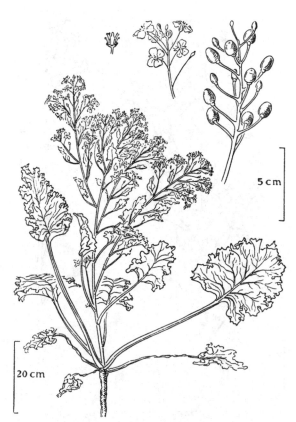

168.　*Crambe maritima* L.　Seakale　White

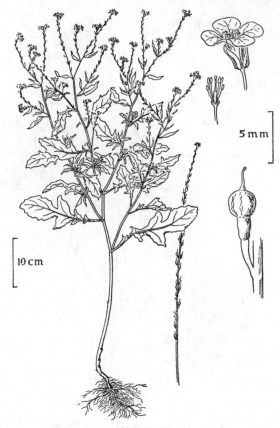

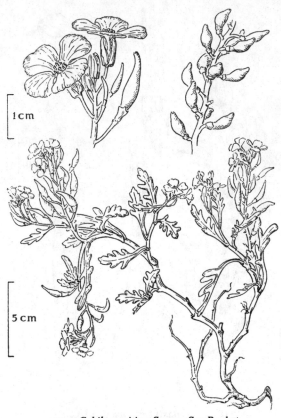

169.   *Rapistrum rugosum* (L.) All.   Yellow

170.   *Cakile maritima* Scop.   Sea Rocket
Purple, lilac or white

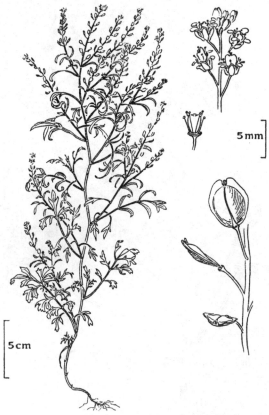

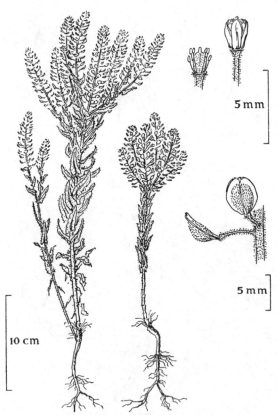

171.   *Lepidium sativum* L.   Garden Cress
White or reddish

172.   *Lepidium campestre* (L.) R.Br.
Pepperwort   White

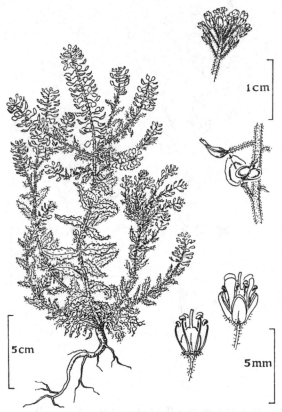

173. *Lepidium heterophyllum* (DC.) Benth. (*L. smithii* Hook.)
'Smith's Cress' White

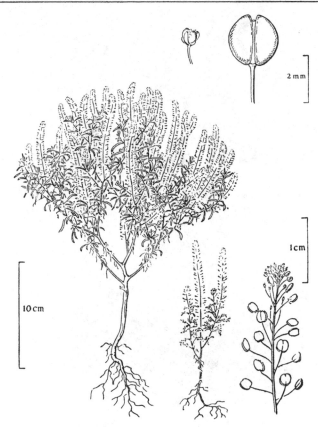

174. *Lepidium ruderale* L. 'Narrow-leaved Pepperwort'
Greenish

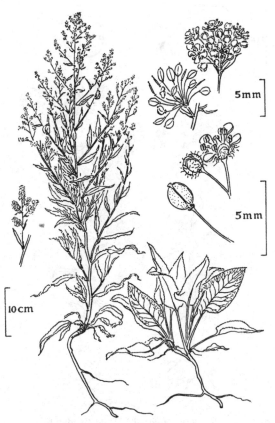

175. *Lepidium latifolium* L. Dittander White

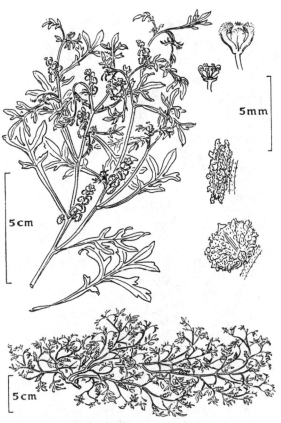

176. *Coronopus squamatus* (Forsk.) Aschers.
Swine-cress, Wart-cress White

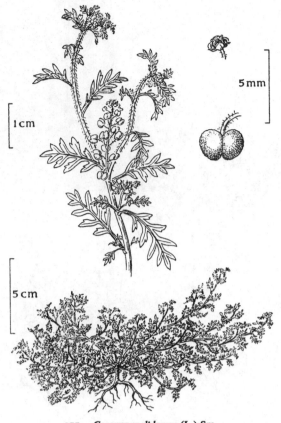

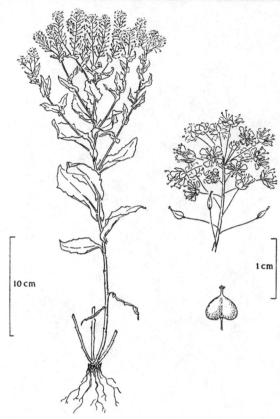

177. *Coronopus didymus* (L.) Sm.
'Lesser Swine-cress'   White

178. *Cardaria draba* (L.) Desv.   'Hoary Cress'   White

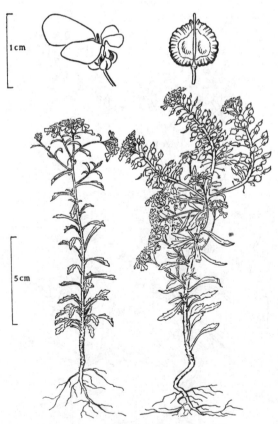

179. *Isatis tinctoria* L.   Woad   Yellow

180. *Iberis amara* L.   Wild Candytuft
White or mauve

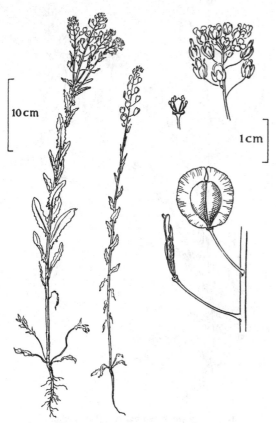

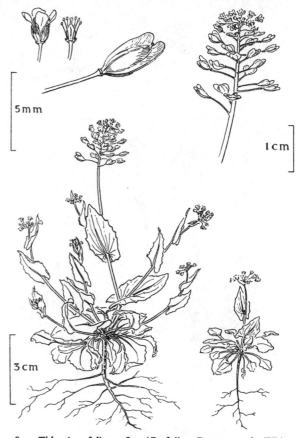

181.   *Thlaspi arvense* L.   Field Penny-cress   White

182.   *Thlaspi perfoliatum* L.   'Perfoliate Penny-cress'   White

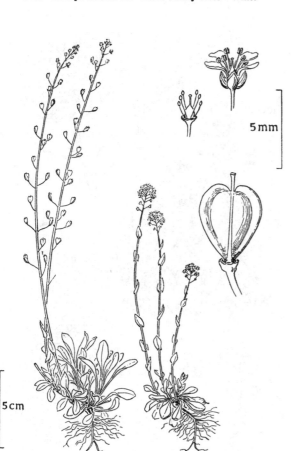

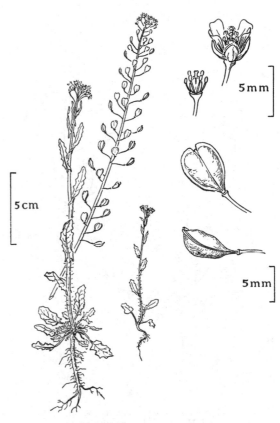

183.   *Thlaspi alpestre* L.   'Alpine Penny-cress'   White

184.   *Thlaspi alliaceum* L.   White

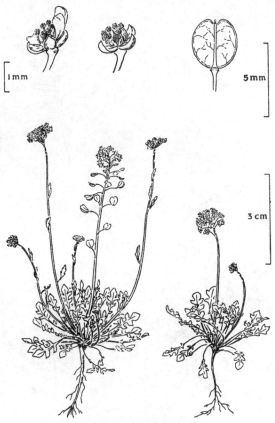

185. *Teesdalia nudicaulis* (L.) R.Br.
'Shepherd's Cress'   White

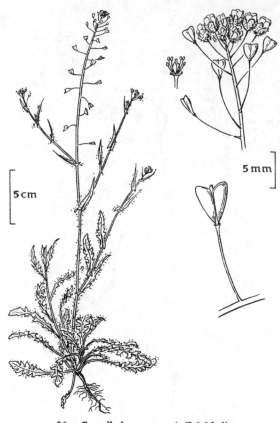

186. *Capsella bursa-pastoris* (L.) Medic.
Shepherd's Purse   White

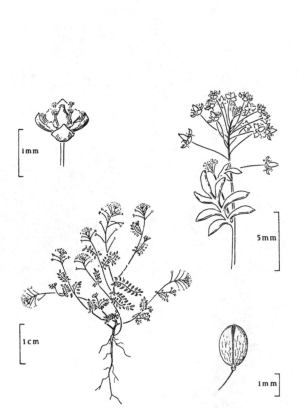

187. *Hornungia petraea* (L.) Rchb.
'Rock Hutchinsia'   Greenish-white

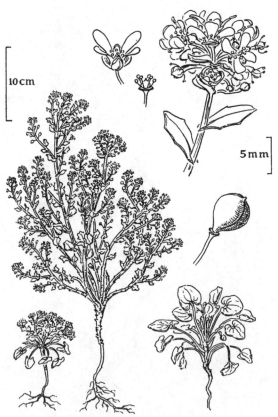

188. *Cochlearia officinalis* L.   Scurvy-grass
White or lilac

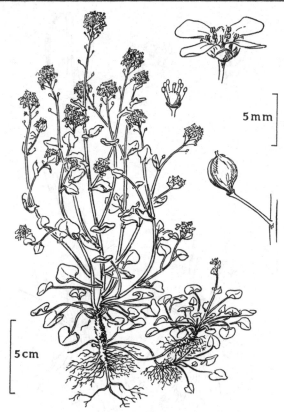

189.   *Cochlearia alpina* (Bab.) H. C. Wats.
'Mountain Scurvy-grass'   White

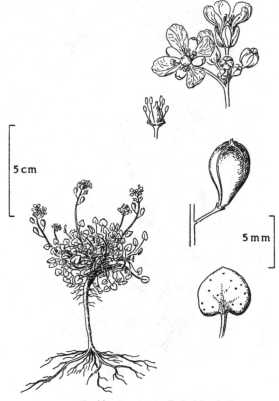

190.   *Cochlearia micacea* E. S. Marshall
'Scottish Scurvy-grass'   White

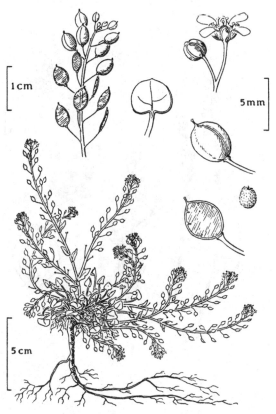

191.   *Cochlearia scotica* Druce
'Scottish Scurvy-grass'   Pale mauve

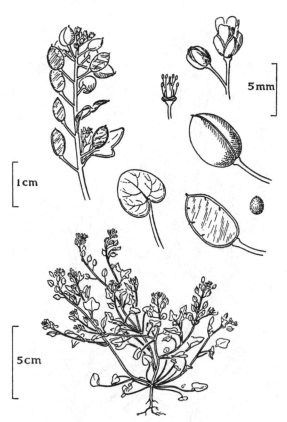

192.   *Cochlearia danica* L.   'Danish Scurvy-grass'
Mauve or whitish

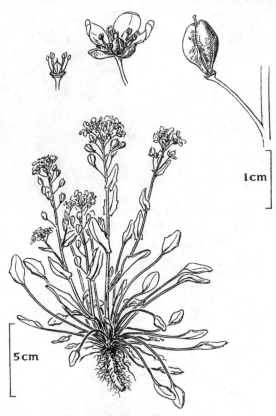

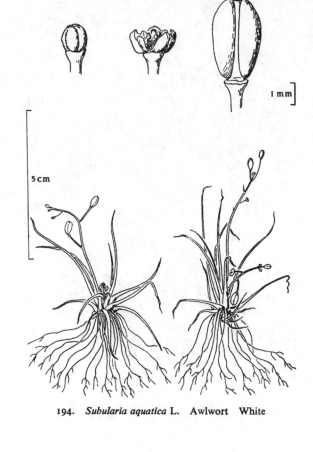

193. *Cochlearia anglica* L. 'Long-leaved Scurvy-grass'
White or pale mauve

194. *Subularia aquatica* L. Awlwort White

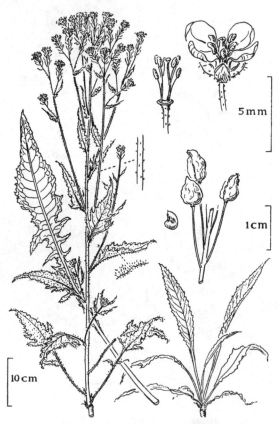

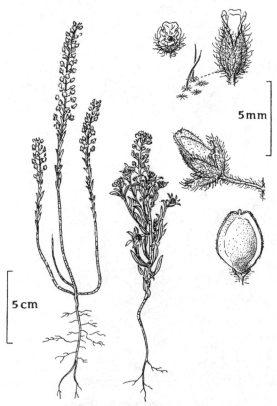

195. *Bunias orientalis* L. Yellow

196. *Alyssum alyssoides* (L.) L. 'Small Alison'
Pale yellow

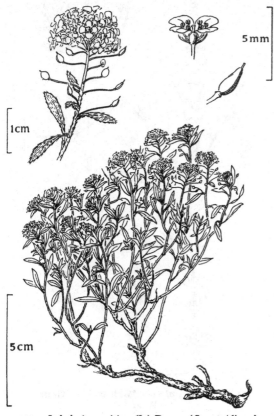

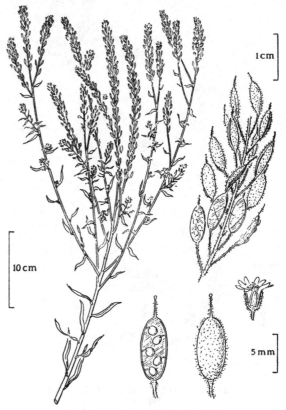

197. *Lobularia maritima* (L.) Desv. 'Sweet Alison'
           White

198. *Berteroa incana* (L.) DC. White

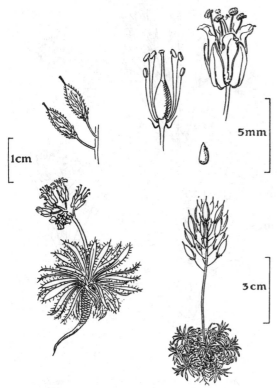

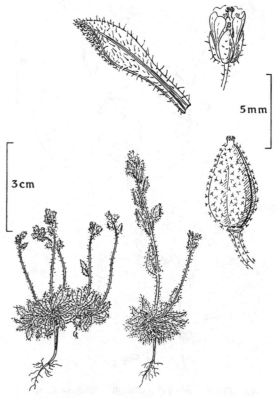

199. *Draba aizoides* L. 'Yellow Whitlow Grass'
           Yellow

200. *Draba norvegica* Gunn. (*D. rupestris* R.Br.)
      'Rock Whitlow Grass' White

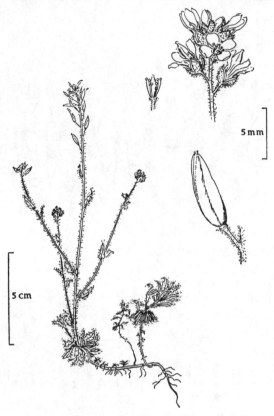

201.   *Draba incana* L.   'Hoary Whitlow Grass'
White

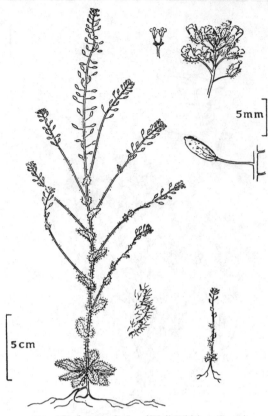

202.   *Draba muralis* L.   'Wall Whitlow Grass'
White

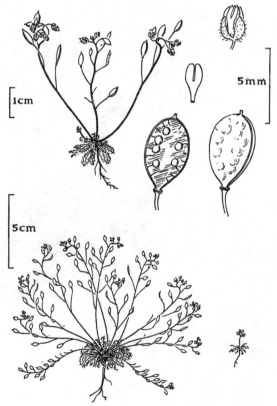

203.   *Erophila verna* (L.) Chevall.   Whitlow Grass
White

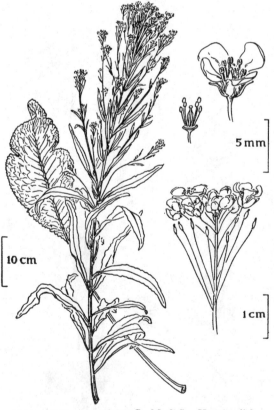

204.   *Armoracia rusticana* G., M. & S.   Horse-radish
White

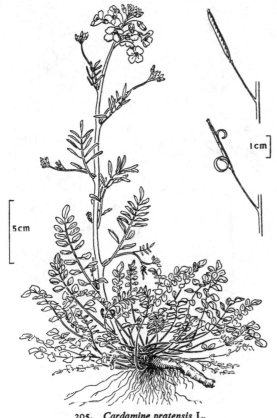

205.   *Cardamine pratensis* L.
Cuckoo Flower, Lady's Smock   Lilac

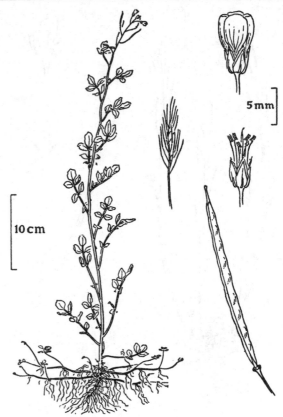

206.   *Cardamine amara* L.   'Large Bitter-cress'   White

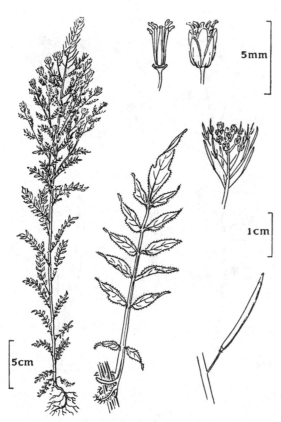

207.   *Cardamine impatiens* L.
'Narrow-leaved Bitter-cress'   White

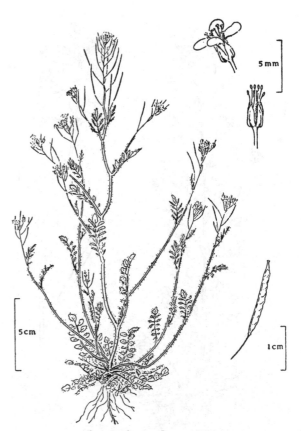

208.   *Cardamine flexuosa* With.
'Wood Bitter-cress'   White

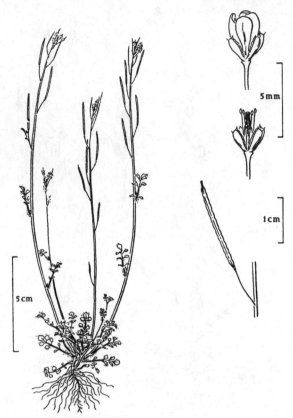

209. *Cardamine hirsuta* L. 'Hairy Bitter-cress' White

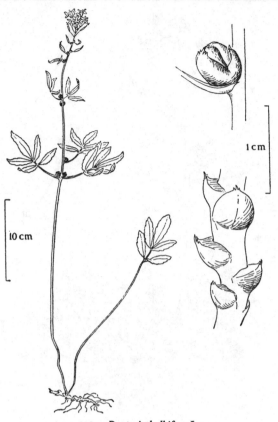

210. *Dentaria bulbifera* L.
Coral-wort. Purple, pink or white

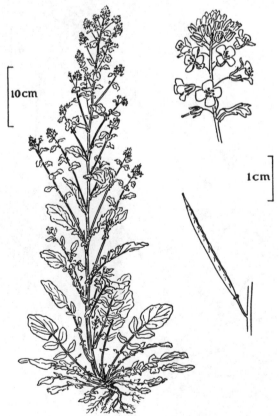

211. *Barbarea vulgaris* R.Br. 'Winter Cress' Yellow

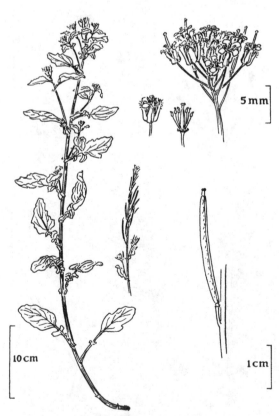

212. *Barbarea stricta* Andrz.
'Small-flowered Yellow Rocket' Yellow

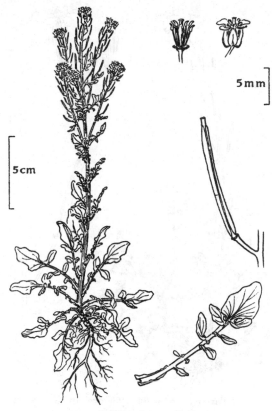

5mm

5cm

213.   *Barbarea intermedia* Bor.
'Intermediate Yellow Rocket'   Yellow

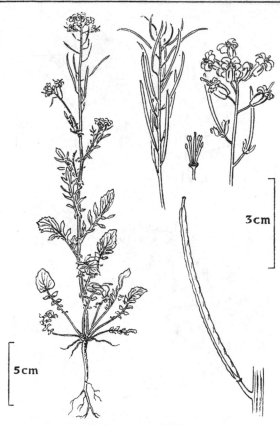

3cm

5cm

214.   *Barbarea verna* (Mill.) Aschers.
Land-cress   Yellow

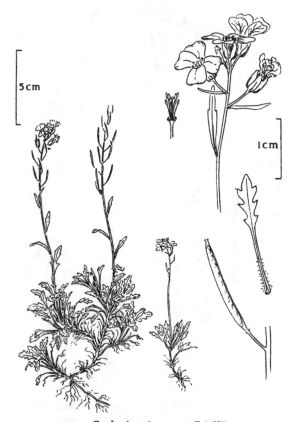

5cm

1cm

215.   *Cardaminopsis petraea* (L.) Hiit.
Northern Rock-cress   White or purplish

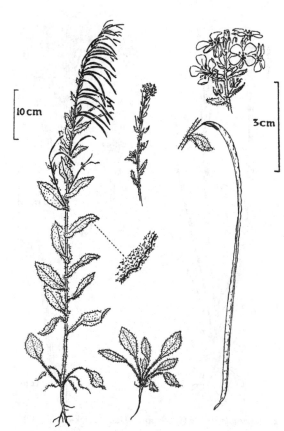

10cm

3cm

216.   *Arabis turrita* L.   'Tower-cress'   Pale yellow

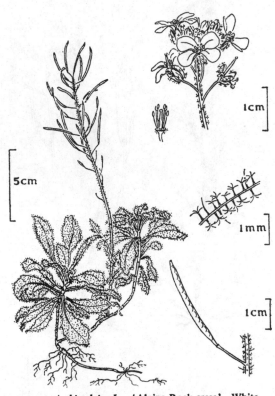

217. *Arabis alpina* L. 'Alpine Rock-cress' White

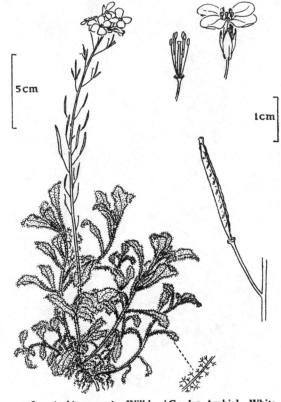

218. *Arabis caucasica* Willd. 'Garden Arabis' White

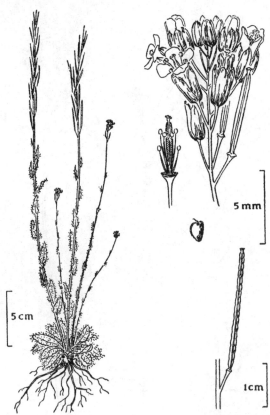

219. *Arabis hirsuta* (L.) Scop. 'Hairy Rock-cress' White

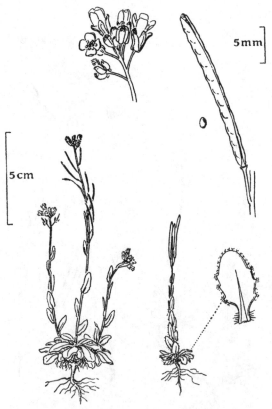

220. *Arabis brownii* Jord. 'Fringed Rock-cress' White

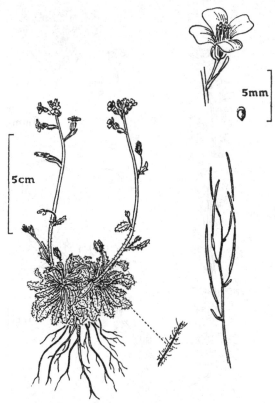

221. *Arabis stricta* Huds. 'Bristol Rock-cress' Cream

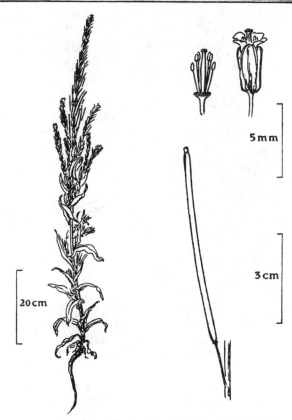

222. *Turritis glabra* L. Tower Mustard Whitish

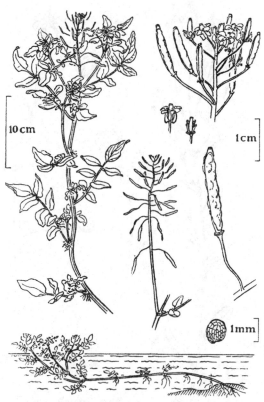

223. *Nasturtium officinale* R.Br. Watercress White

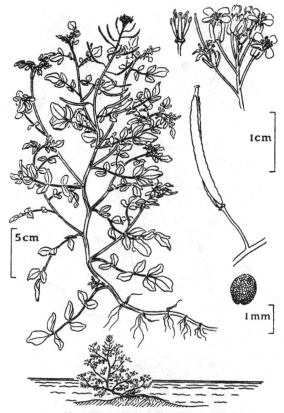

224. *Nasturtium microphyllum* (Boenn.) Rchb.
'One-rowed Watercress' White

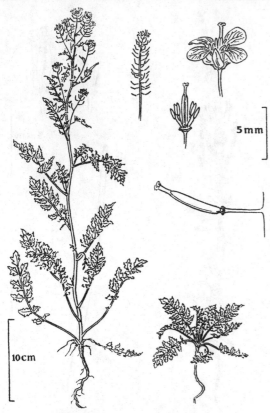

225. *Rorippa sylvestris* (L.) Bess.
'Creeping Yellow-cress'   Yellow

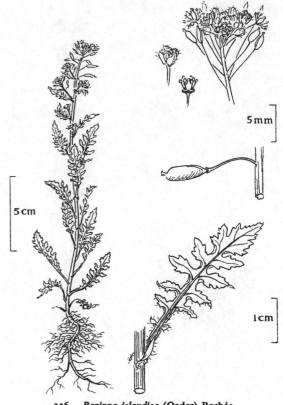

226. *Rorippa islandica* (Oeder) Borbás
'Marsh Yellow-cress'   Yellow

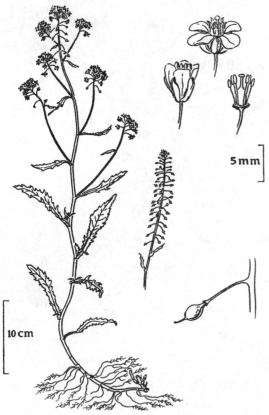

227. *Rorippa amphibia* (L.) Bess.
'Great Yellow-cress'   Yellow

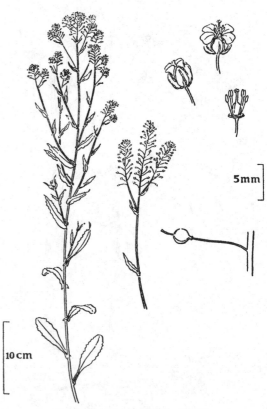

228. *Rorippa austriaca* (Crantz) Bess.
'Austrian Yellow-cress'   Yellow

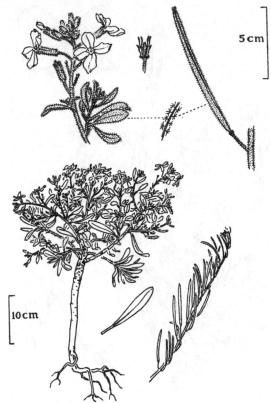

229. *Matthiola incana* (L.) R.Br.
Stock, Gilliflower   Purple, red or white

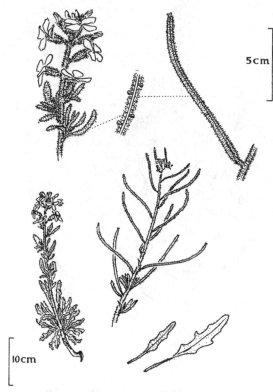

230. *Matthiola sinuata* (L.) R.Br.
'Sea Stock'   Pale purple

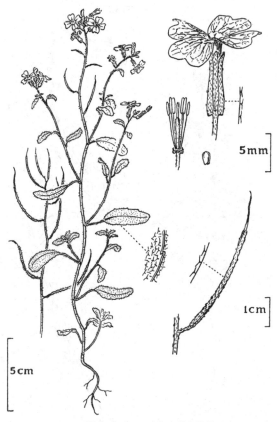

231. *Malcolmia maritima* (L.) R.Br.
Virginia Stock   Violet, pink or white

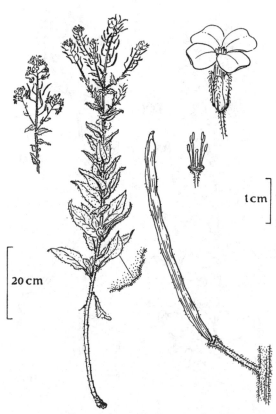

232. *Hesperis matronalis* L.   Dame's Violet
Violet or white

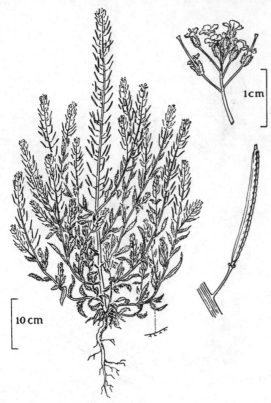

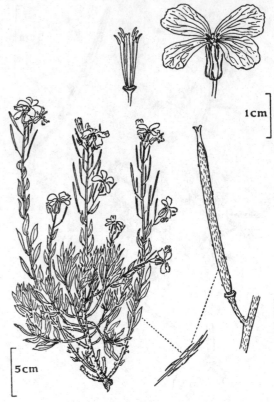

233. *Erysimum cheiranthoides* L.   Treacle Mustard
        Yellow

234.   *Cheiranthus cheiri* L.   Wallflower
         Orange-yellow

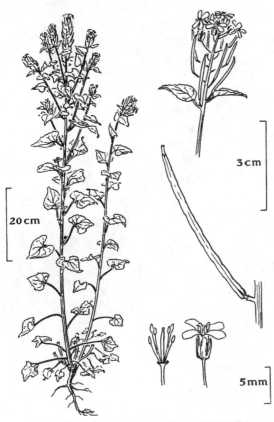

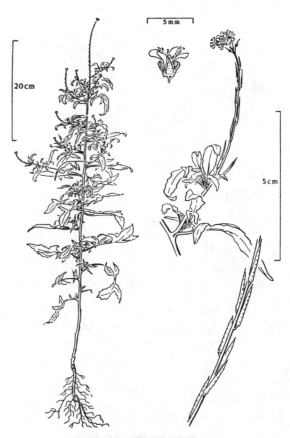

235. *Alliaria petiolata* (Bieb.) Cavara & Grande
Garlic Mustard, Hedge Garlic, Jack-by-the-Hedge   White

236. *Sisymbrium officinale* (L.) Scop.
         Hedge Mustard   Yellow

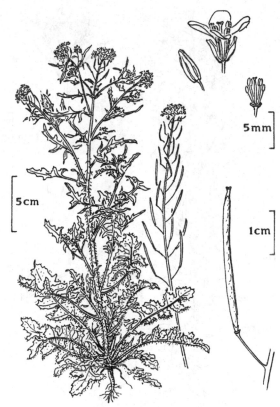

237.   *Sisymbrium loeselii* L.   Yellow

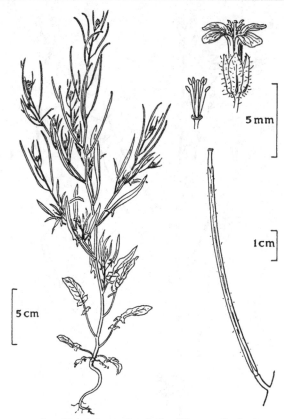

238.   *Sisymbrium orientale* L.   'Eastern Rocket'
Pale yellow

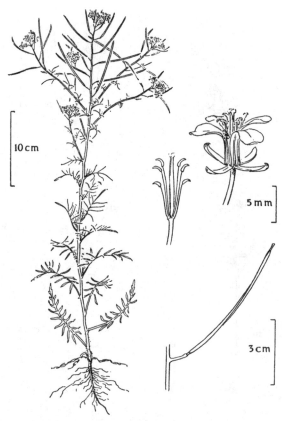

239.   *Sisymbrium altissimum* L.   'Tall Rocket'
Pale yellow

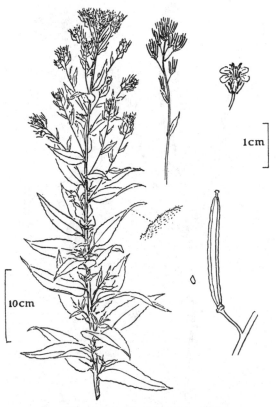

240.   *Sisymbrium strictissimum* L.   Yellow

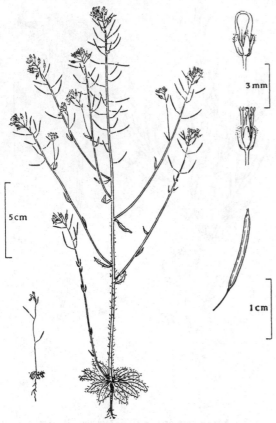

241.　*Arabidopsis thaliana* (L.) Heynh.　Thale Cress
　　　White

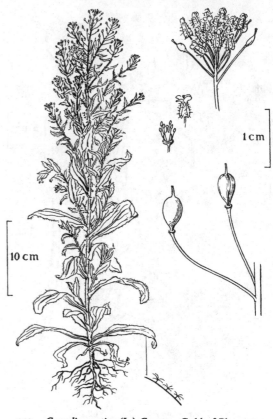

242.　*Camelina sativa* (L.) Crantz　Gold of Pleasure
　　　Yellow

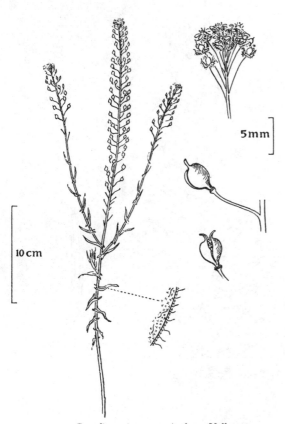

243.　*Camelina microcarpa* Andrz.　Yellow

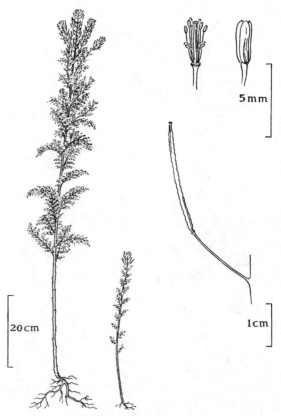

244.　*Descurainia sophia* (L.) Webb　Flixweed　Yellow

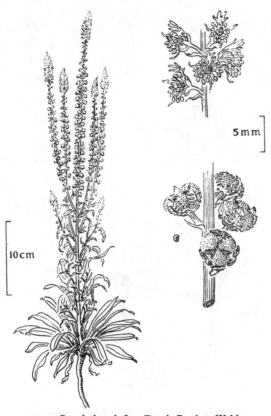

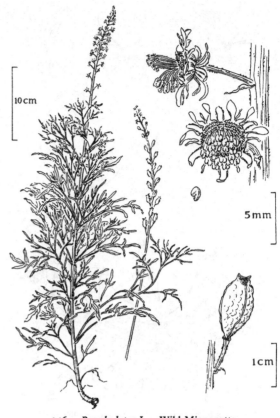

245.   *Reseda luteola* L.   Dyer's Rocket, Weld
Yellow-green

246.   *Reseda lutea* L.   Wild Mignonette
Yellow-green

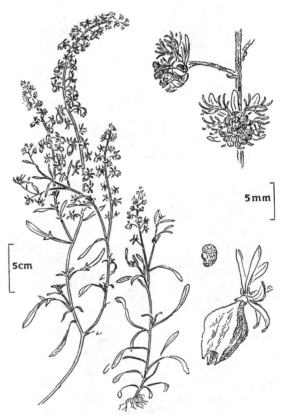

247.   *Reseda alba* L.   'Upright Mignonette'   Whitish

248.   *Reseda phyteuma* L.   Whitish

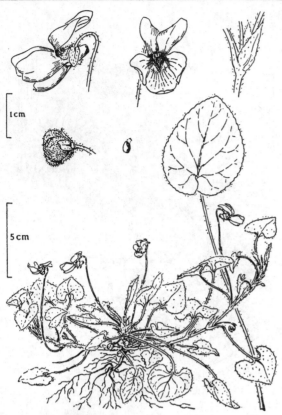

249. *Viola odorata* L. Sweet Violet   Violet or white

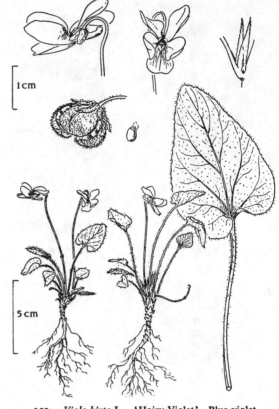

250. *Viola hirta* L. 'Hairy Violet'   Blue-violet

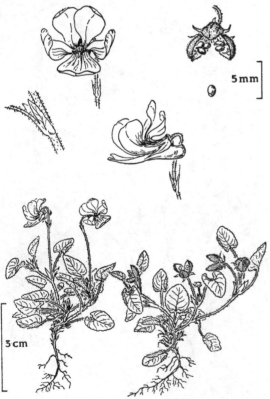

251. *Viola rupestris* Schmidt 'Teesdale Violet'
Blue-violet

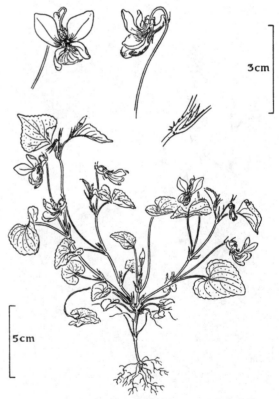

252. *Viola riviniana* Rchb. 'Common Violet'
Blue-violet

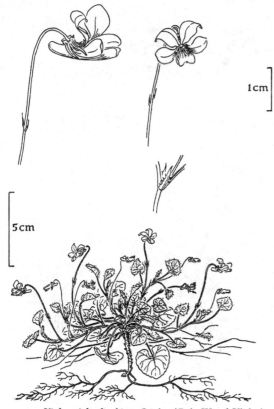

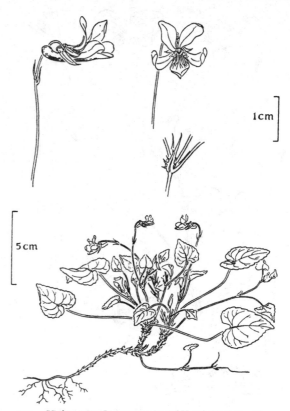

253.   *Viola reichenbachiana* Jord.   'Pale Wood Violet'
Lilac

254.   *Viola canina* L. ssp. *canina*   'Heath Violet'   Blue

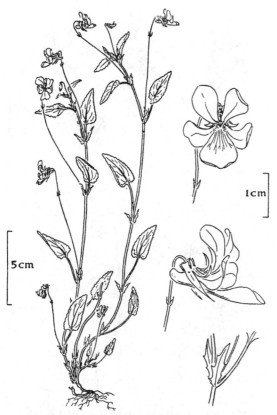

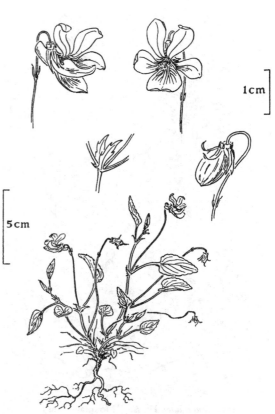

255.   *Viola canina* ssp. *montana* (L.) Fr.   Pale blue

256.   *Viola lactea* Sm.   'Pale Heath Violet'
Pale greyish violet

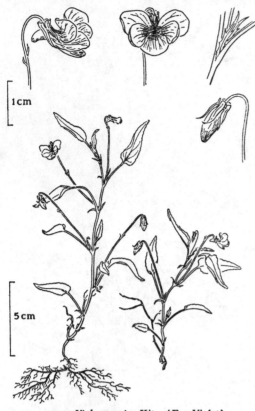

257. *Viola stagnina* Kit. 'Fen Violet'
Bluish white or white

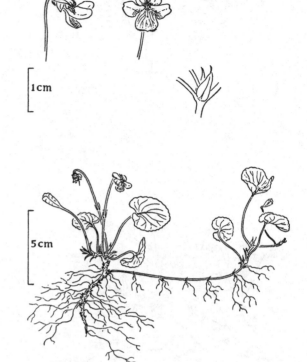

258. *Viola palustris* L. 'Marsh violet' Lilac

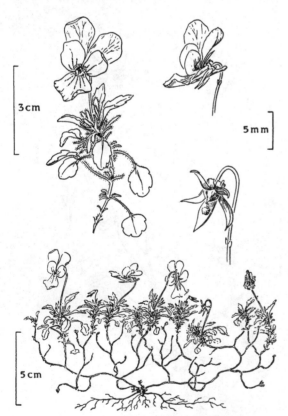

259. *Viola lutea* Huds. 'Mountain Pansy'
Yellow, violet

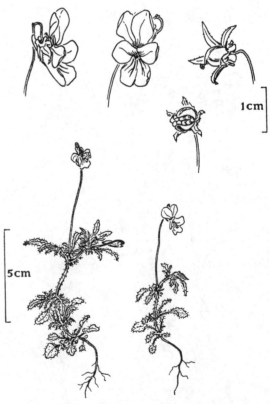

260. *Viola tricolor* L. ssp. *tricolor* Wild Pansy
Pale yellow, violet

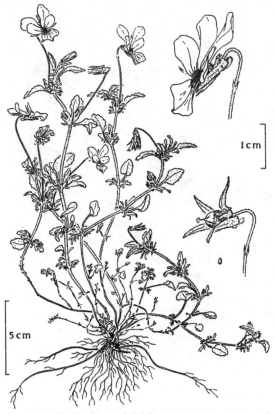

261.   *Viola tricolor* ssp. *saxatilis* (Schmidt) Rouy & Foucaud
Pale yellow, violet

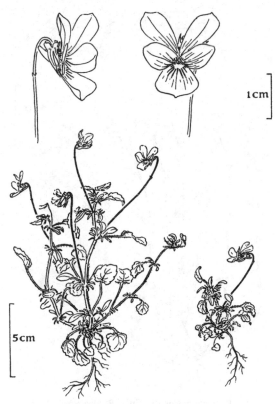

262.   *Viola tricolor* ssp. *curtisii* (E. Forst.) Syme
Pale yellow, violet

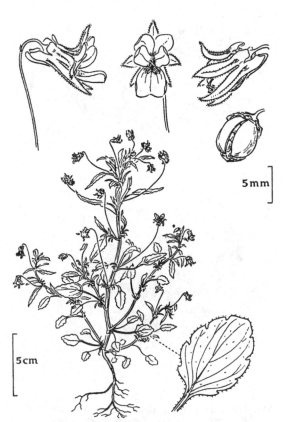

263.   *Viola arvensis* Murr.   'Field Pansy'
Cream or violet-tinged

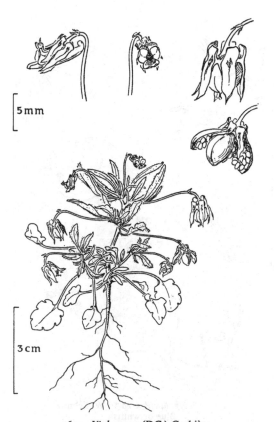

264.   *Viola nana* (DC.) Corbière
Cream or violet-tinged

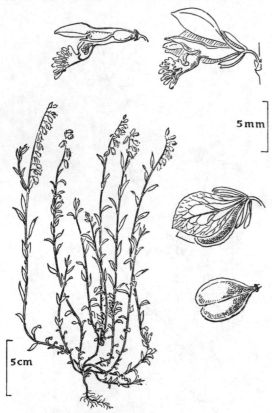

265.   *Polygala vulgaris* L. (incl. *P. oxyptera* Rchb.)
'Common Milkwort'   Blue, pink or white

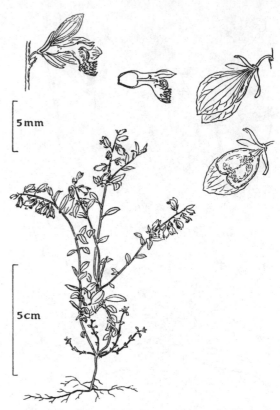

266.   *Polygala serpyllifolia* Hose 'Common Milkwort'
Blue, pink or white

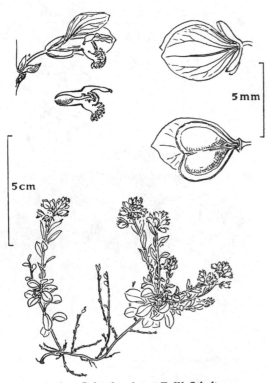

267.   *Polygala calcarea* F. W. Schultz
Blue or whitish

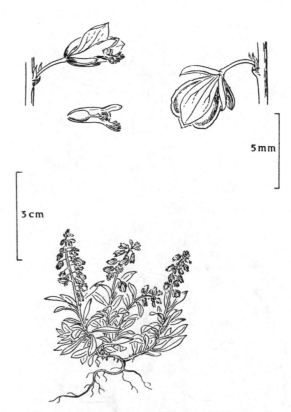

268.   *Polygala amara* L. (incl. *P. amarella* Crantz and
P. austriaca Crantz) Pink, purplish or blue

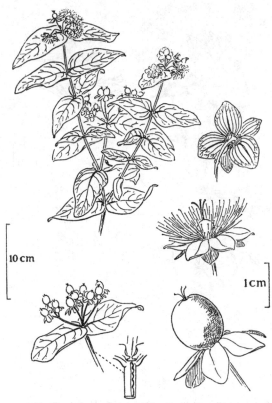

269. *Hypericum androsaemum* L.   Tutsan   Yellow

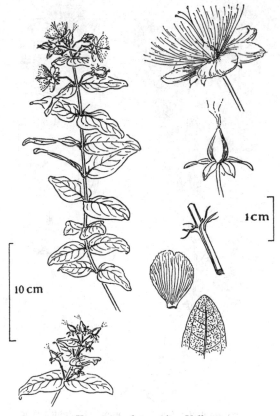

270. *Hypericum elatum* Ait.   Yellow

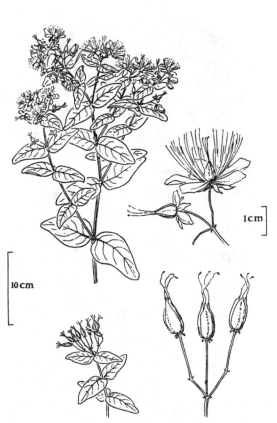

271. *Hypericum hircinum* L.   Yellow

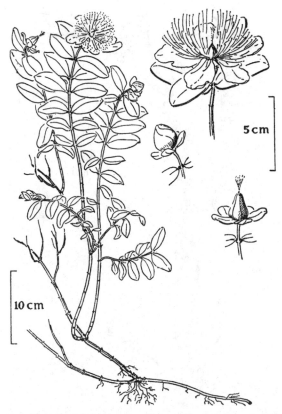

272. *Hypericum calycinum* L.   Rose of Sharon, Aaron's Beard
        Yellow

[69]

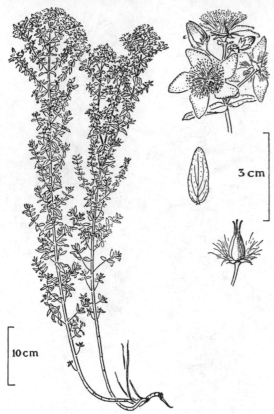

273. *Hypericum perforatum* L. 'Common St John's Wort'
            Yellow

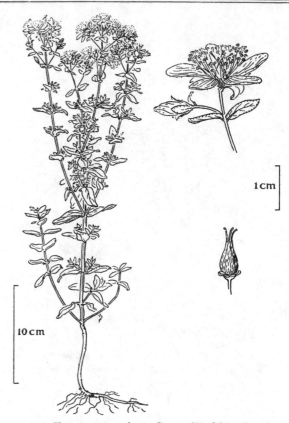

274. *Hypericum maculatum* Crantz (*H. dubium* Leers)
        'Imperforate St John's Wort'   Yellow

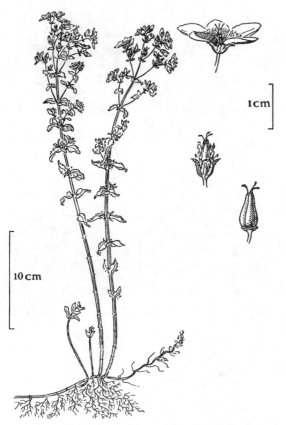

275. *Hypericum undulatum* Schousb.   Yellow

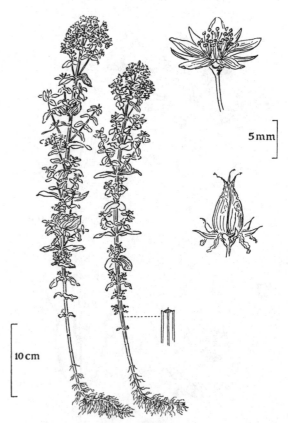

276. *Hypericum tetrapterum* Fr.
      'Square-stemmed St John's Wort'   Yellow

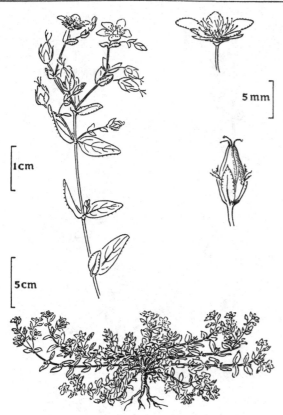

5 mm

1cm

5cm

277.   *Hypericum humifusum* L.
'Trailing St John's Wort'   Yellow

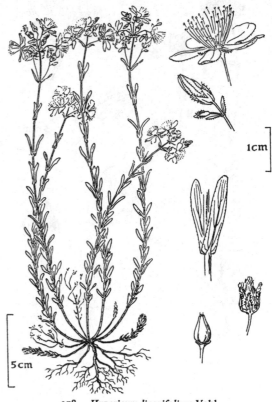

1cm

5cm

278.   *Hypericum linarifolium* Vahl
'Flax-leaved St John's Wort'   Yellow

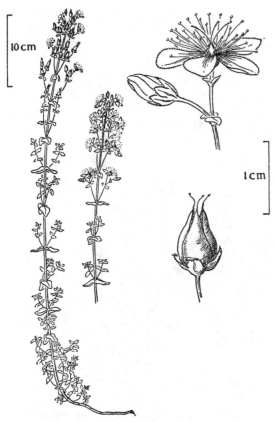

10cm

1cm

279.   *Hypericum pulchrum* L.
'Slender St John's Wort'   Yellow

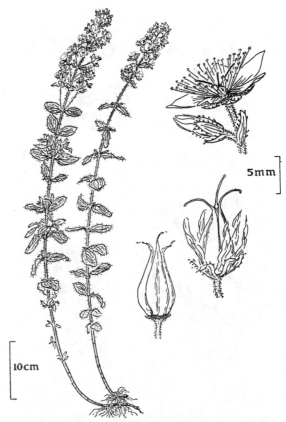

5mm

10cm

280.   *Hypericum hirsutum* L.   'Hairy St John's Wort'
Yellow

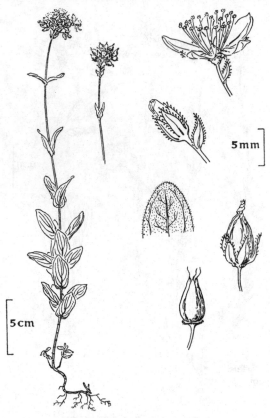

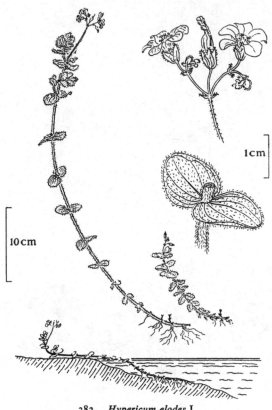

281.   *Hypericum montanum* L.
'Mountain St John's Wort'   Yellow

282.   *Hypericum elodes* L.
'Marsh St John's Wort'   Yellow

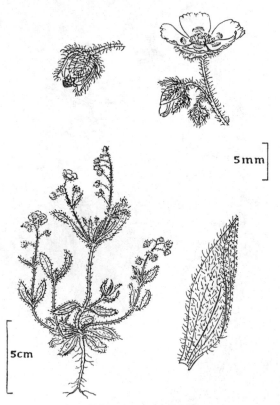

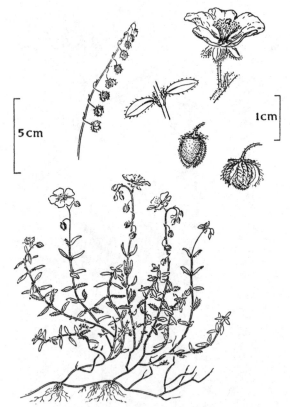

283.   *Tuberaria guttata* (L.) Fourr.
(*Helianthemum guttatum* (L.) Mill.)   'Annual Rockrose'
Pale yellow

284.   *Helianthemum chamaecistus* Mill.   Common Rockrose
Yellow

285. *Helianthemum apenninum* (L.) Mill.
White Rockrose   White

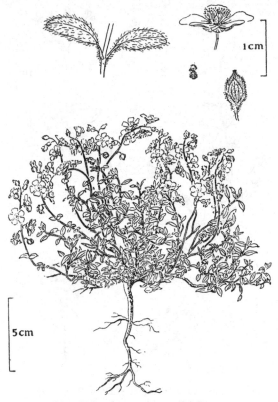

286. *Helianthemum canum* (L.) Baumg.
'Hoary Rockrose'   Yellow

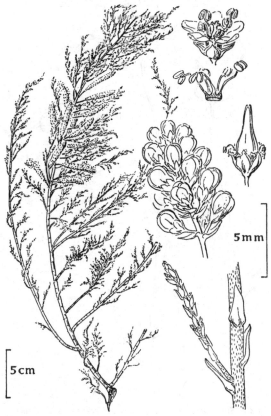

287. *Tamarix anglica* Webb   Tamarisk   Pink or white

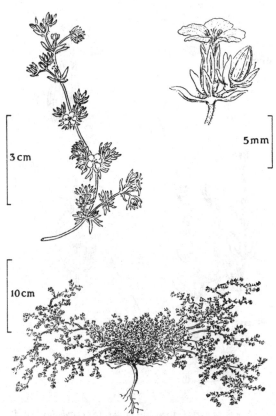

288. *Frankenia laevis* L.   'Sea Heath'   Pink

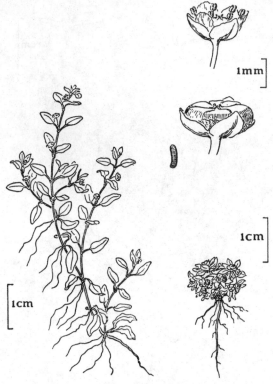

289. *Elatine hexandra* (Lapierre) DC. Pinkish white

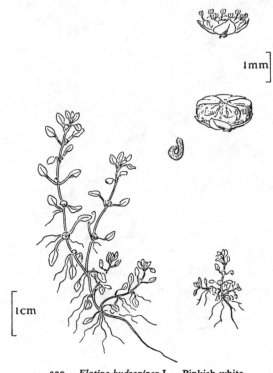

290. *Elatine hydropiper* L. Pinkish white

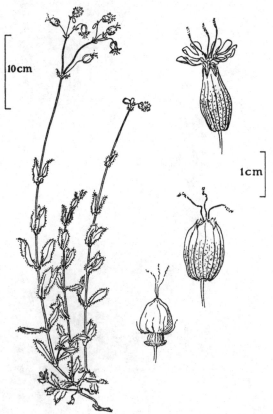

291. *Silene vulgaris* (Moench) Garcke (*S. cucubalus* Wibel)
Bladder Campion White

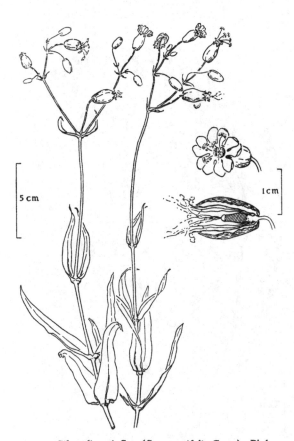

292. *Silene linearis* Sw. (*S. angustifolia* Guss.) Pink

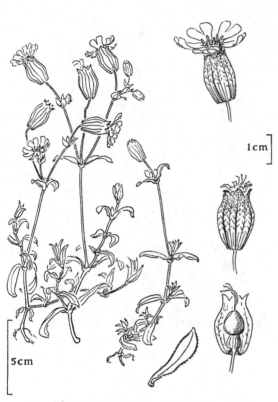

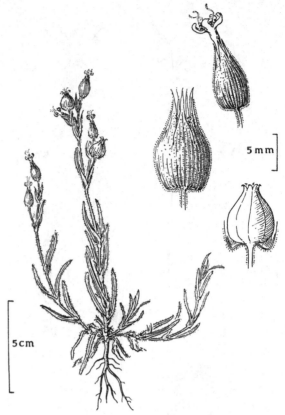

293. *Silene maritima* With.   Sea Campion   White

294. *Silene conica* L.   'Striated Catchfly'   Red

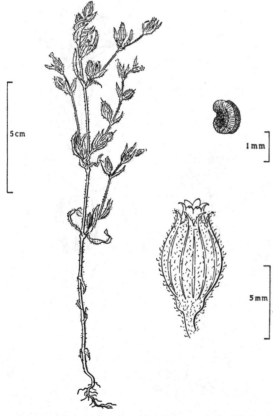

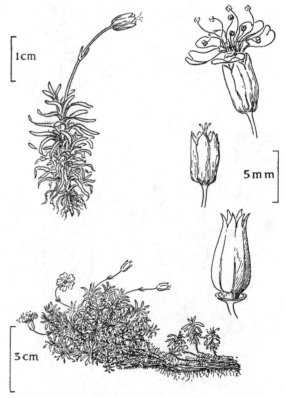

295. *Silene anglica* L.   'Small-flowered Catchfly'
White or pink

296. *Silene acaulis* (L.) Jacq.   Moss Campion
Pink or whitish

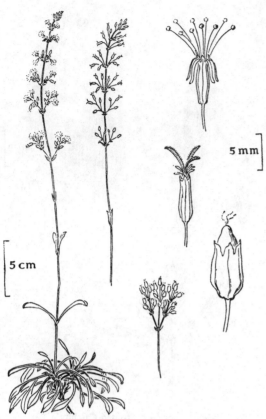

297. *Silene otites* (L.) Wibel   Spanish Catchfly
Yellowish green

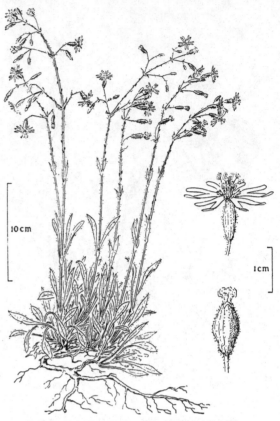

298. *Silene nutans* L.   Nottingham Catchfly
White or pinkish

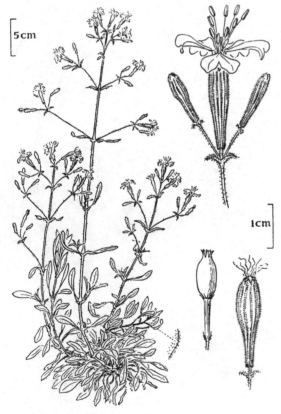

299. *Silene italica* (L.) Pers.   'Italian Catchfly'
Yellowish white

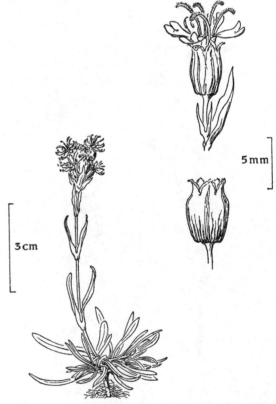

300. *Viscaria alpina* (L.) Don   'Red Alpine Catchfly'
Pink

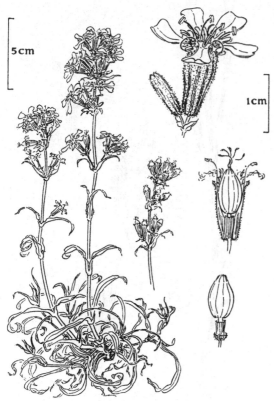

301. *Viscaria vulgaris* Bernh.
'Red German Catchfly'   Red

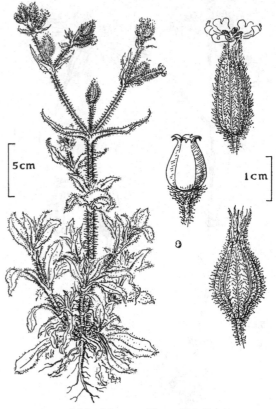

302. *Melandrium noctiflorum* (L.) Garcke
'Night-flowering Campion'   Pinkish yellow

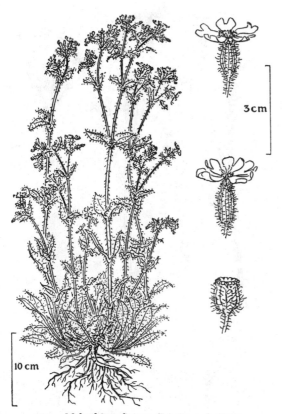

303. *Melandrium dioicum* (L.) Coss. & Germ.
(*M. rubrum* (Weigel) Garcke)   Red Campion   Red

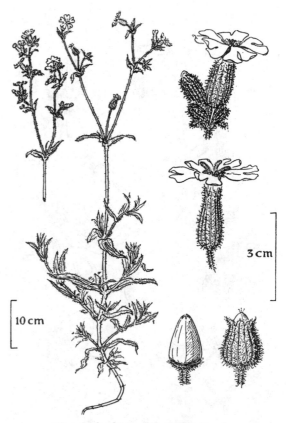

304. *Melandrium album* (Mill.) Garcke
White Campion   White

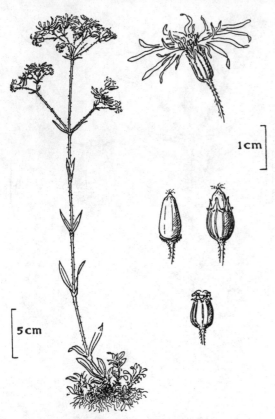

305.   *Lychnis flos-cuculi* L.   Ragged Robin   Red

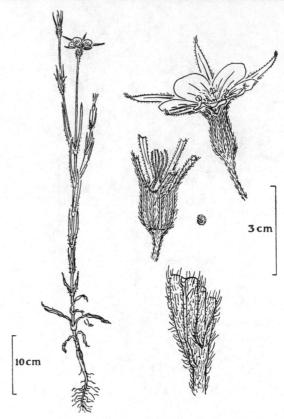

306.   *Agrostemma githago* L.   Corn Cockle   Purple

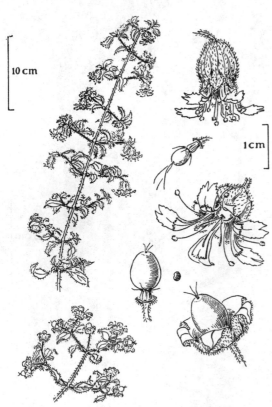

307.   *Cucubalus baccifer* L.   'Berry Catchfly'
           Greenish white

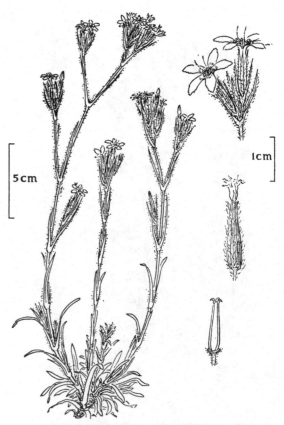

308.   *Dianthus armeria* L.   Deptford Pink   Red

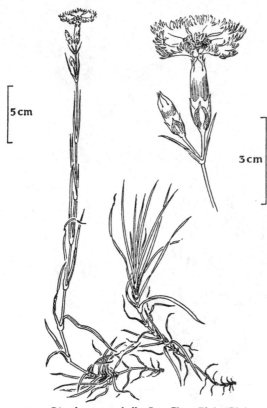

309. *Dianthus caryophyllus* L.   Clove Pink Pink

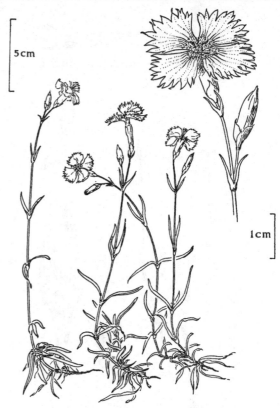

310. *Dianthus gratianopolitanus* Vill.   Cheddar Pink Pink

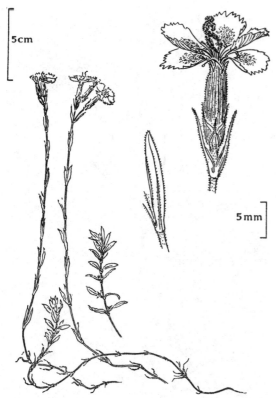

311. *Dianthus deltoides* L.   Maiden Pink Pink

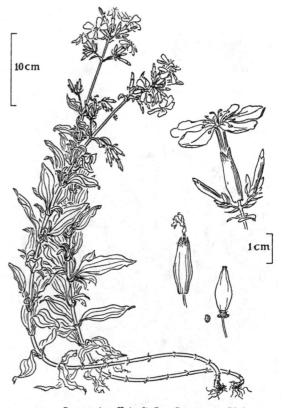

312. *Saponaria officinalis* L.   Soapwort Pink

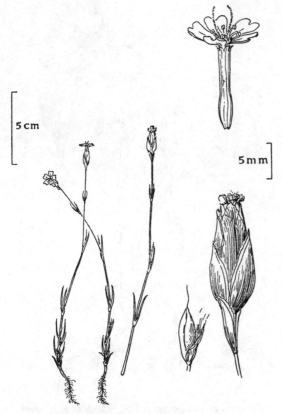

313. *Kohlrauschia prolifera* (L.) Kunth
'Proliferous Pink'   Reddish

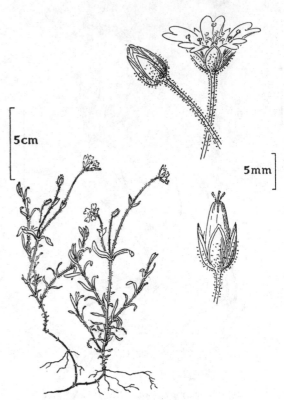

314. *Cerastium cerastoides* (L.) Britton
'Starwort Mouse-ear Chickweed'   White

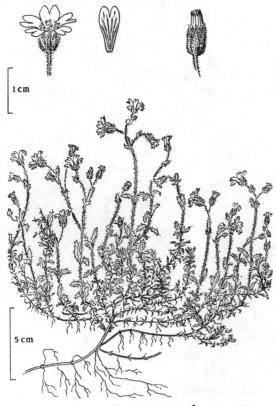

315. *Cerastium arvense* L.
'Field Mouse-ear Chickweed'   White

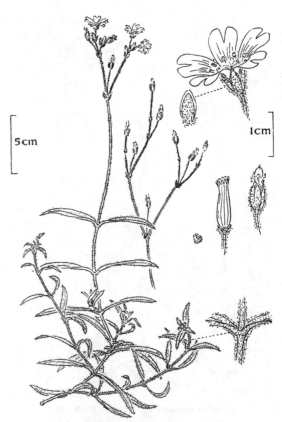

316. *Cerastium tomentosum* L.
Dusty Miller, Snow-in-Summer   White

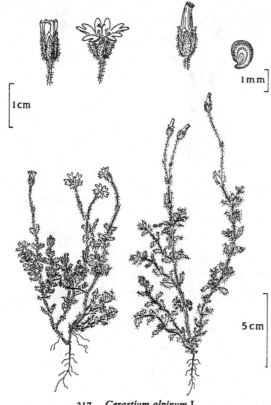

317. *Cerastium alpinum* L.
'Alpine Mouse-ear Chickweed' White

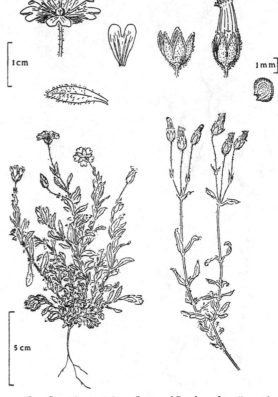

318. *Cerastium arcticum* Lange (*C. edmondstonii* auct.)
'Arctic Mouse-ear Chickweed' White

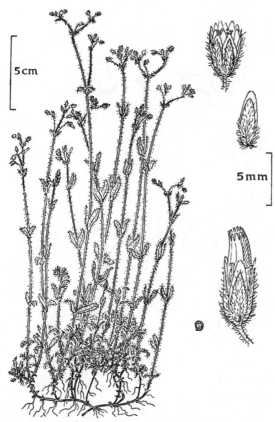

319. *Cerastium holosteoides* Fr. (*C. vulgatum* auct.)
'Common Mouse-ear Chickweed' White

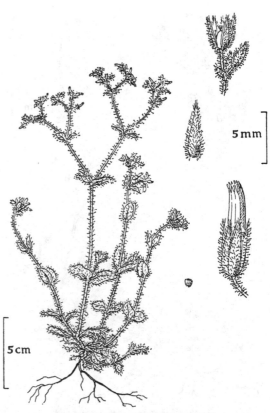

320. *Cerastium glomeratum* Thuill.
'Sticky Mouse-ear Chickweed' White

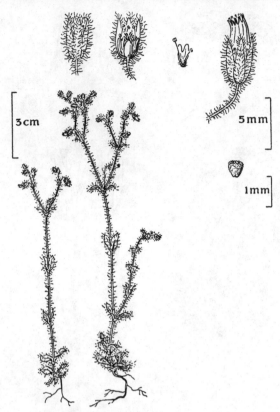

321.   *Cerastium brachypetalum* Pers.   White

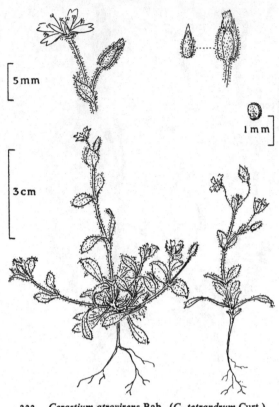

322.   *Cerastium atrovirens* Bab. (*C. tetrandrum* Curt.)
'Dark-green Mouse-ear Chickweed'   White

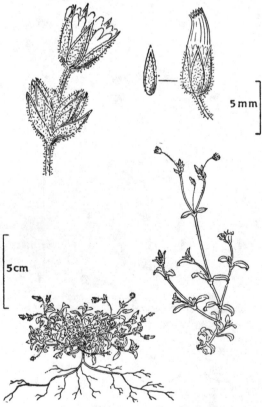

323.   *Cerastium pumilum* Curt.
'Curtis's Mouse-ear Chickweed'   White

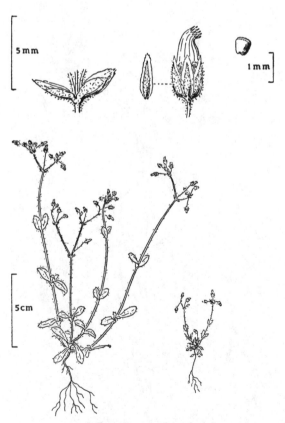

324.   *Cerastium semidecandrum* L.
'Little Mouse-ear Chickweed'   White

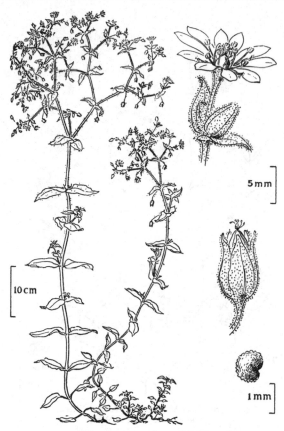

325. *Myosoton aquaticum* (L.) Moench
Water Chickweed   White

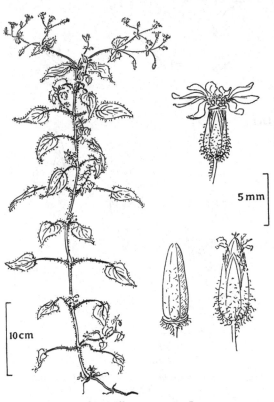

326. *Stellaria nemorum* L.
Wood Stitchwort, Wood Chickweed   White

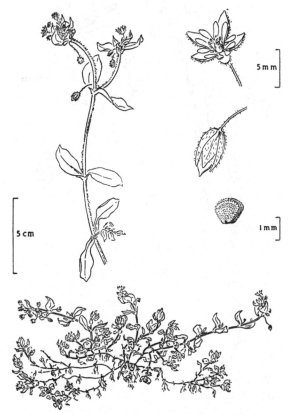

327. *Stellaria media* (L.) Vill.   Chickweed   White

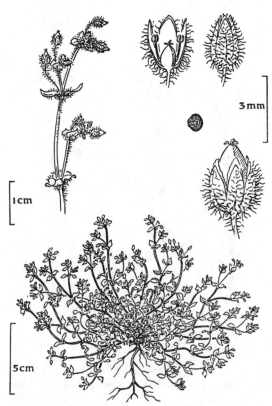

328. *Stellaria pallida* (Dumort.) Piré (*S. apetala* auct.)
'Lesser Chickweed'   White

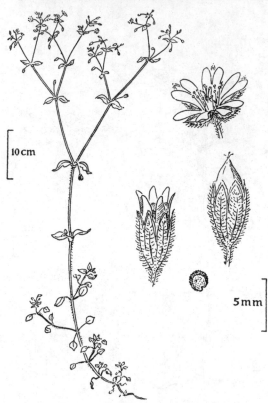

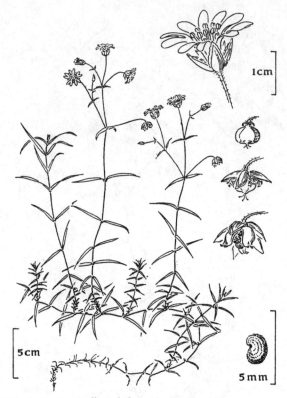

329. *Stellaria neglecta* Weihe   'Greater Chickweed'
White

330. *Stellaria holostea* L.   Adders' Meat,
'Greater Stitchwort'   White

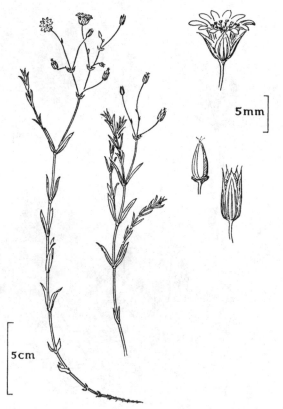

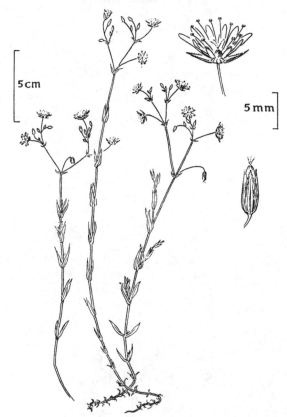

331. *Stellaria palustris* Retz. 'Marsh Stitchwort' White

332. *Stellaria graminea* L. 'Lesser Stitchwort' White

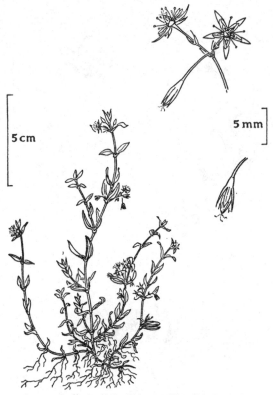

333.   *Stellaria alsine* Grimm   'Bog Stitchwort'
White

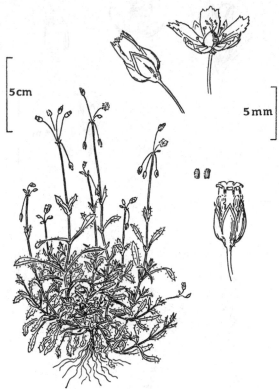

334.   *Holosteum umbellatum* L.   'Jagged Chickweed'
White

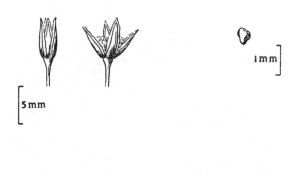

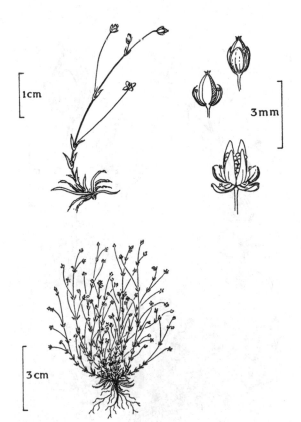

335.   *Moenchia erecta* (L.) G., M. & S.
'Upright Chickweed'   White

336.   *Sagina apetala* Ard.   'Common Pearlwort'
Greenish

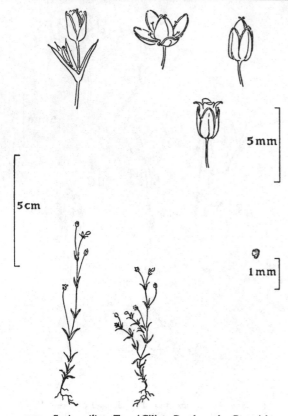

337.   *Sagina ciliata* Fr.   'Ciliate Pearlwort'   Greenish

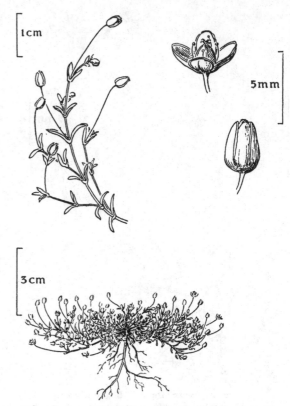

338.   *Sagina maritima* Don   'Sea Pearlwort'   Greenish

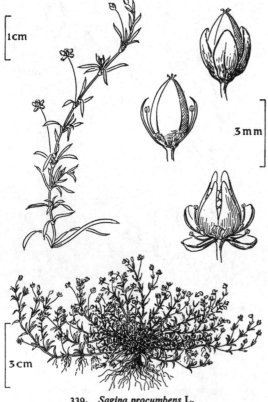

339.   *Sagina procumbens* L.
Procumbent Pearlwort'   Greenish

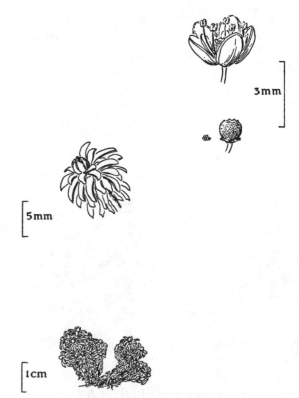

340.   *Sagina boydii* F. B. White
'Boyd's Pearlwort'   Greenish

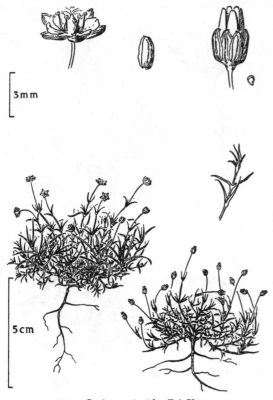

341. *Sagina saginoides* (L.) Karst.
'Alpine Pearlwort' White

342. *Sagina intermedia* Fenzl
'Lesser Alpine Pearlwort' White

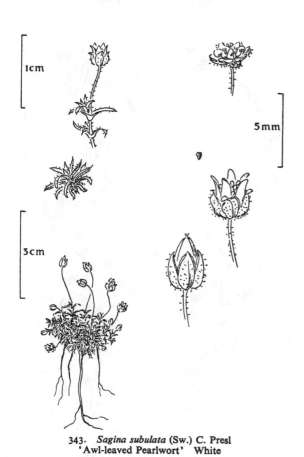

343. *Sagina subulata* (Sw.) C. Presl
'Awl-leaved Pearlwort' White

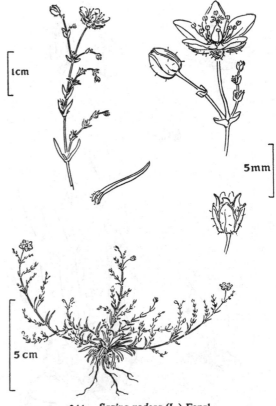

344. *Sagina nodosa* (L.) Fenzl
'Knotted Pearlwort' White

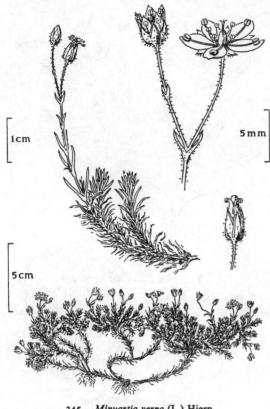

345. *Minuartia verna* (L.) Hiern
'Vernal Sandwort' White

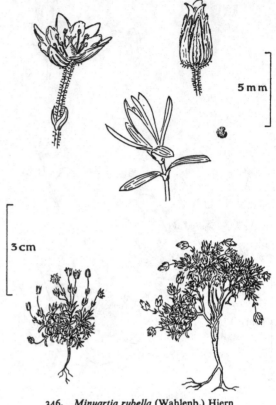

346. *Minuartia rubella* (Wahlenb.) Hiern
'Alpine Sandwort' White

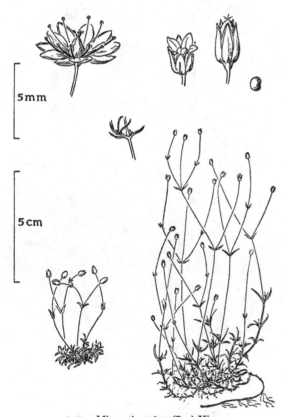

347. *Minuartia stricta* (Sw.) Hiern
'Bog Sandwort' White

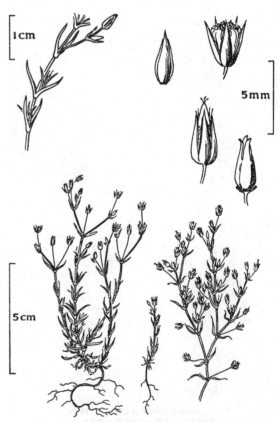

348. *Minuartia tenuifolia* (L.) Hiern
'Fine-leaved Sandwort' White

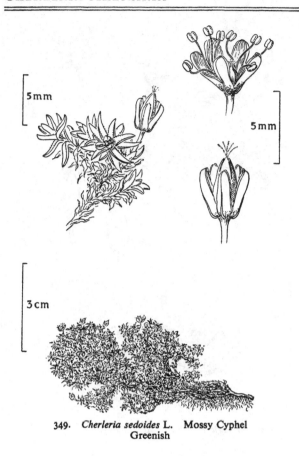

349.   *Cherleria sedoides* L.   Mossy Cyphel
Greenish

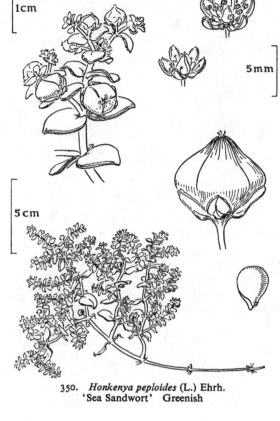

350.   *Honkenya peploides* (L.) Ehrh.
'Sea Sandwort'   Greenish

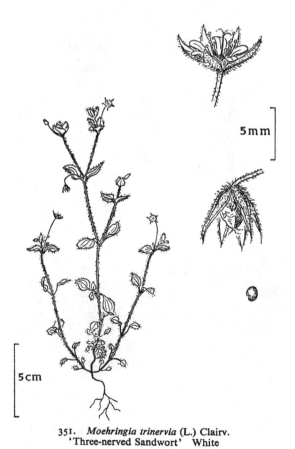

351.   *Moehringia trinervia* (L.) Clairv.
'Three-nerved Sandwort'   White

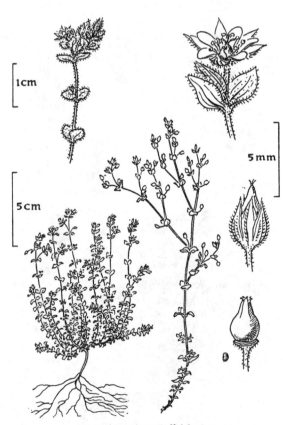

352.   *Arenaria serpyllifolia* L.
'Thyme-leaved Sandwort'   White

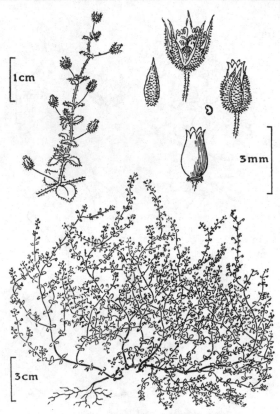

353.   *Arenaria leptoclados* (Rchb.) Guss.
'Lesser Thyme-leaved Sandwort'   White

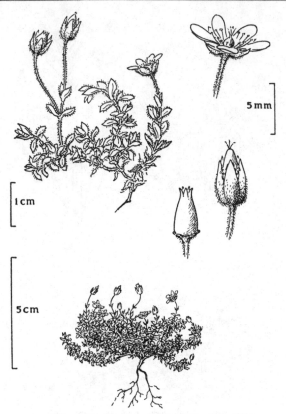

354.   *Arenaria ciliata* L.   'Irish Sandwort'   White

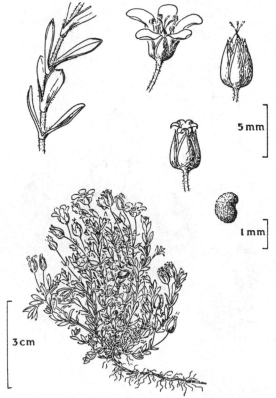

355.   *Arenaria norvegica* Gunn.
'Norwegian Sandwort'   White

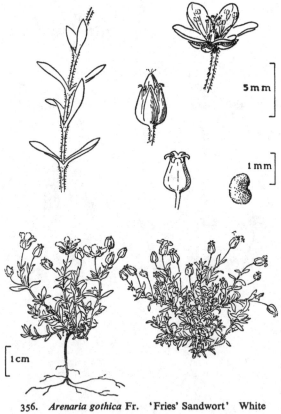

356.   *Arenaria gothica* Fr.   'Fries' Sandwort'   White

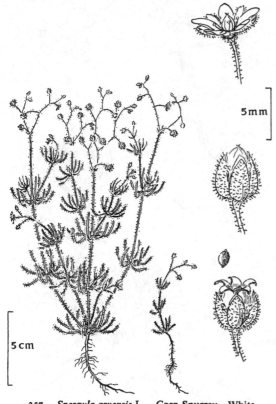

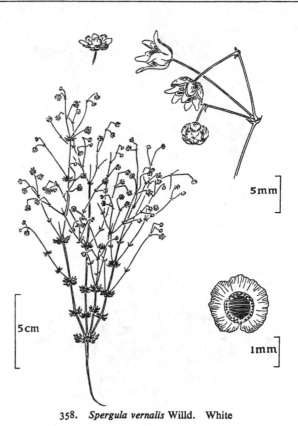

357. *Spergula arvensis* L.   Corn Spurrey   White

358. *Spergula vernalis* Willd.   White

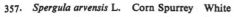

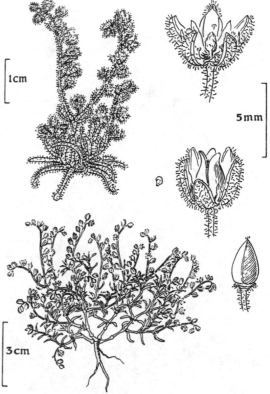

359.   *Spergularia rubra* (L.) J. & C. Presl
Sand-spurrey   Pink

360.   *Spergularia bocconi* (Scheele) Fouc.
'Bocconi's Sand-spurrey'   Pinkish

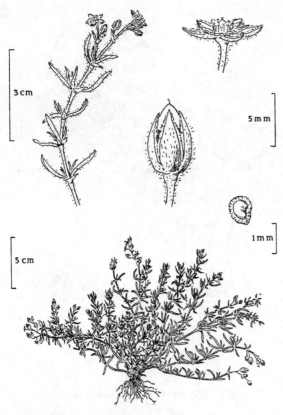

361.   *Spergularia rupicola* Lebel   'Cliff Sand-spurrey'   Pink

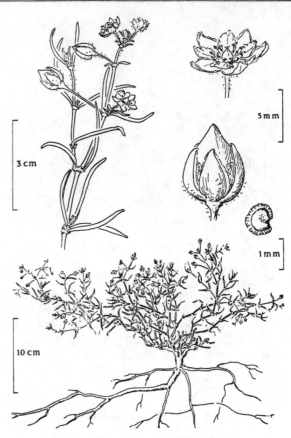

362.   *Spergularia media* (L.) C. Presl
(*S. marginata* Kittel)   Whitish

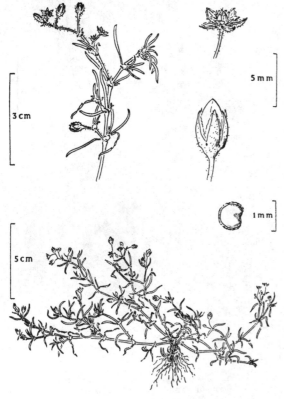

363.   *Spergularia marina* (L.) Griseb.
(*S. salina* J. & C. Presl)   Pink

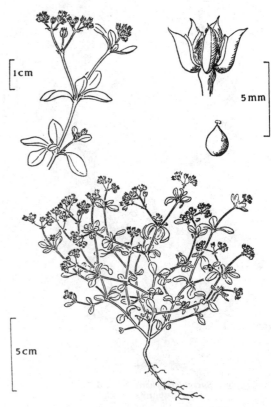

364.   *Polycarpon tetraphyllum* (L.) L.
'Four-leaved All-seed'   Whitish

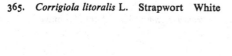

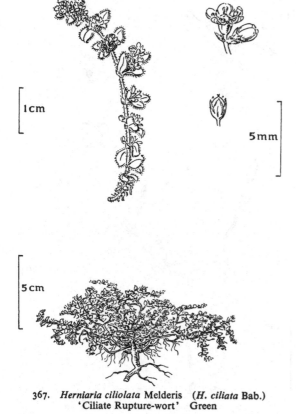

365. *Corrigiola litoralis* L.   Strapwort   White

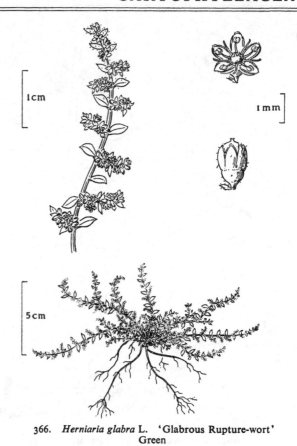

366. *Herniaria glabra* L.   'Glabrous Rupture-wort'
       Green

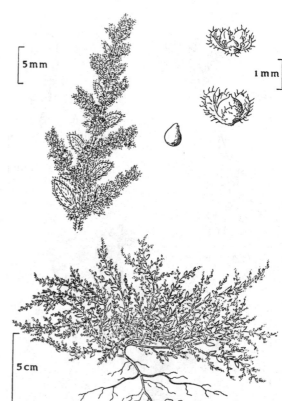

367. *Herniaria ciliolata* Melderis   (*H. ciliata* Bab.)
    'Ciliate Rupture-wort'   Green

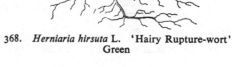

368. *Herniaria hirsuta* L.   'Hairy Rupture-wort'
       Green

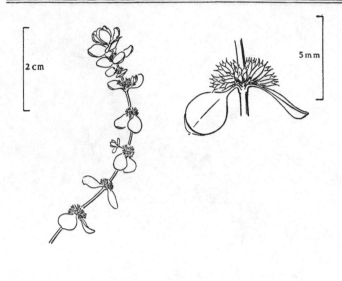

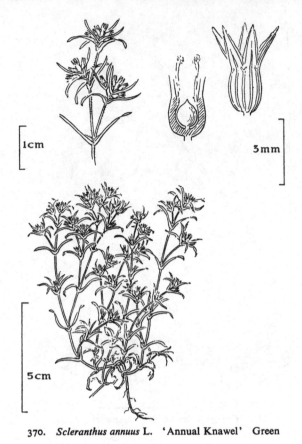

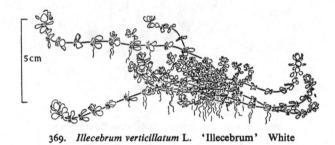

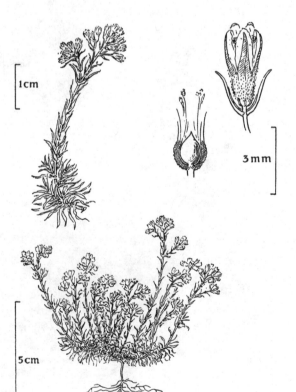

369. *Illecebrum verticillatum* L.   'Illecebrum'   White

370. *Scleranthus annuus* L.   'Annual Knawel'   Green

371. *Scleranthus perennis* L.   'Perennial Knawel'
Green

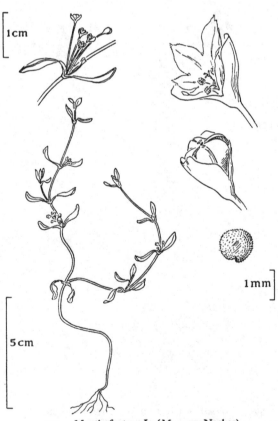

372. *Montia fontana* L. (*M. verna* Necker)
Blinks   Green

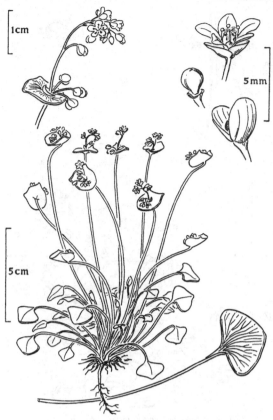

373. *Claytonia perfoliata* Donn ex Willd. White

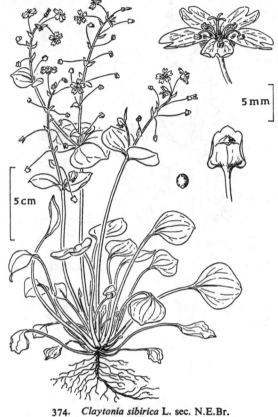

374. *Claytonia sibirica* L. sec. N.E.Br.
(*C. alsinoides* Sims) Pink or white

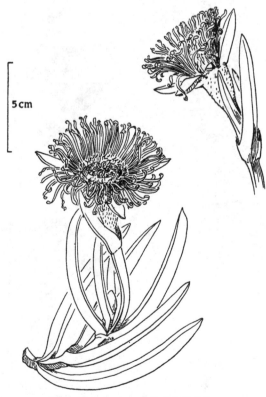

375. *Carpobrotus edulis* (L.) N.E.Br.
Hottentot Fig. Magenta or yellow

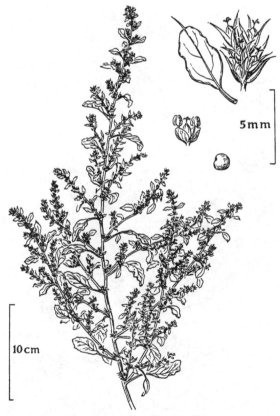

376. *Amaranthus albus* L. Green

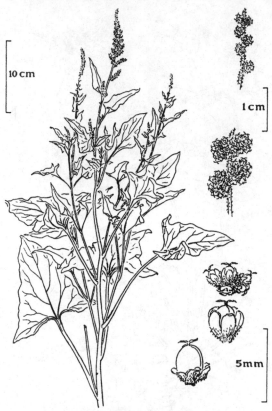

377. *Chenopodium bonus-henricus* L.
Mercury, Good King Henry   Green

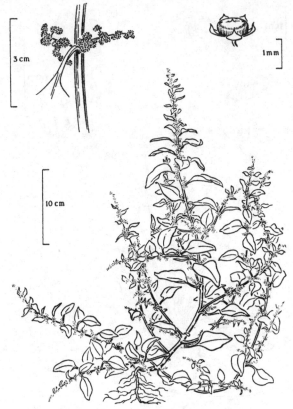

378. *Chenopodium polyspermum* L.   All-seed   Green

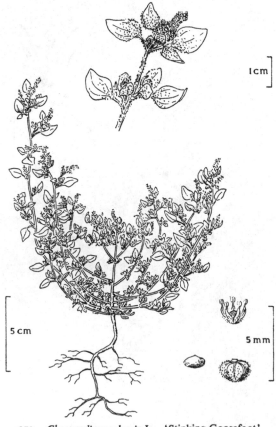

379. *Chenopodium vulvaria* L.   'Stinking Goosefoot'
Green

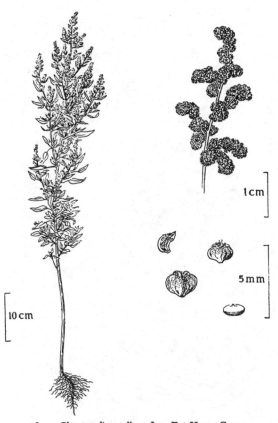

380. *Chenopodium album* L.   Fat Hen   Green

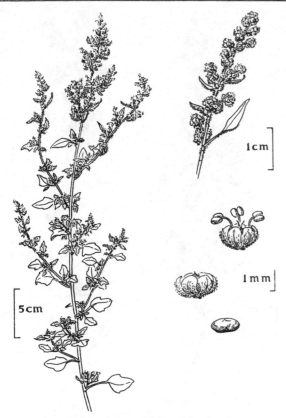

381.   *Chenopodium opulifolium* Schrad.   Green

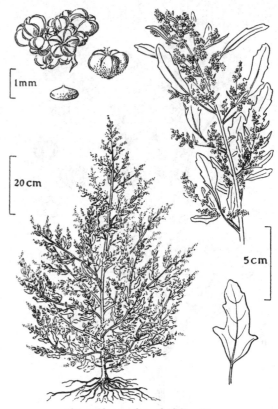

382.   *Chenopodium ficifolium* Sm.
'Fig-leaved Goosefoot'   Green

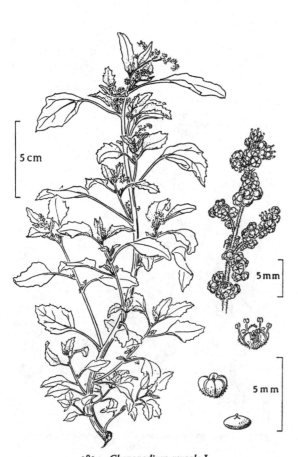

383.   *Chenopodium murale* L.
'Nettle-leaved Goosefoot'   Green

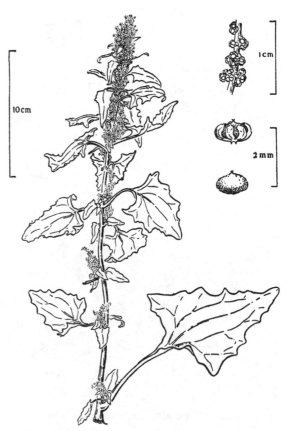

384.   *Chenopodium urbicum* L.   'Upright Goosefoot'
Green

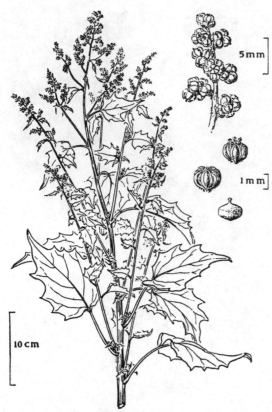

385. *Chenopodium hybridum* L.   Sowbane   Green

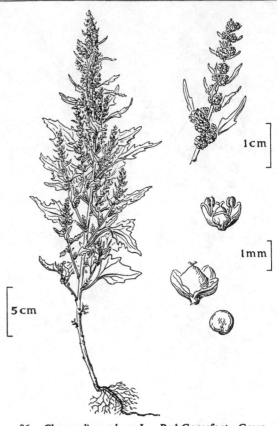

386. *Chenopodium rubrum* L.   Red Goosefoot   Green

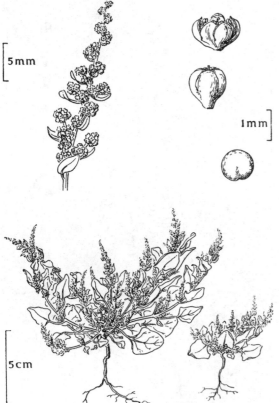

387. *Chenopodium botryodes* Sm.   Green

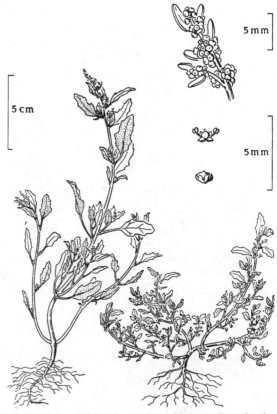

388. *Chenopodium glaucum* L.   'Glaucous Goosefoot'
Green

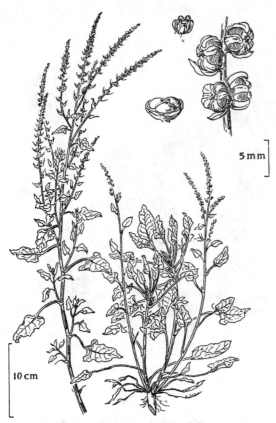

389. *Beta vulgaris* L. ssp. *maritima* (L.) Thell.
Beet   Green

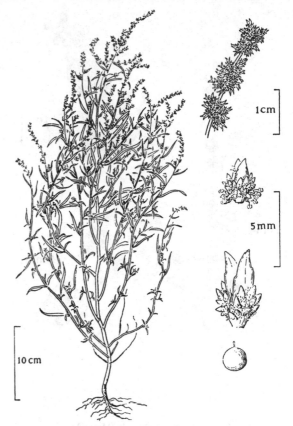

390. *Atriplex littoralis* L. 'Shore Orache'   Green

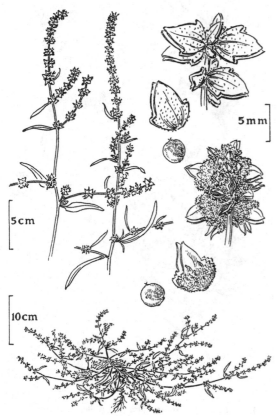

391. *Atriplex patula* L. 'Common Orache'   Green

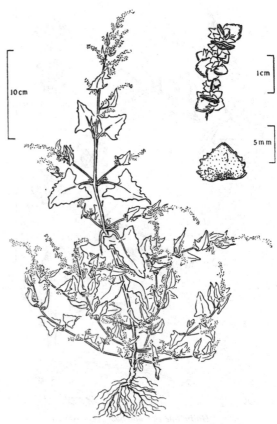

392. *Atriplex hastata* L. 'Hastate Orache'   Green

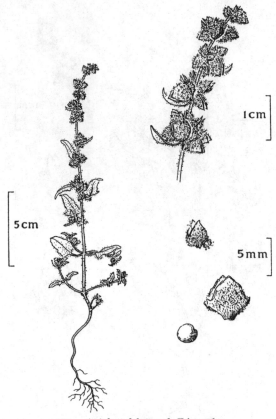

393. *Atriplex glabriuscula* Edmondst.
'Babington's Orache'   Green

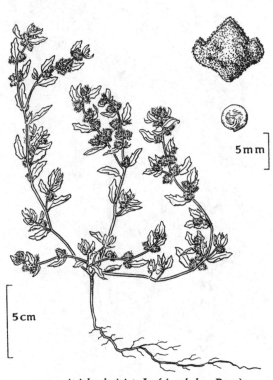

394. *Atriplex laciniata* L. (*A. sabulosa* Rouy)
'Frosted Orache'   Green

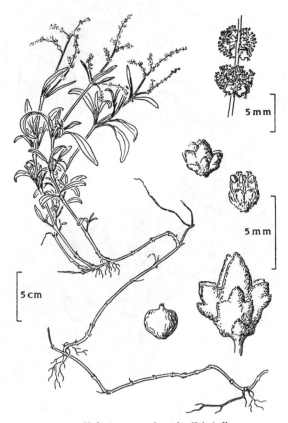

395. *Halimione portulacoides* (L.) Aellen
'Sea Purslane'   Greenish

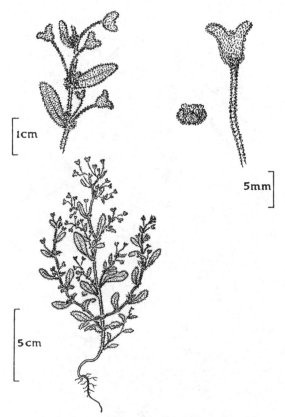

396. *Halimione pedunculata* (L.) Aellen   Greenish

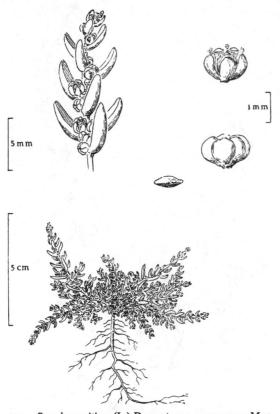

397. *Suaeda maritima* (L.) Dumort. var. *macrocarpa* Moq.
'Herbaceous Seablite'   Green

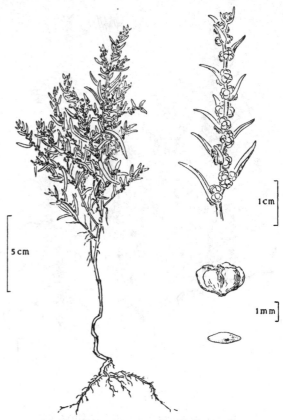

398. *Suaeda maritima* var. *flexilis* Rouy   Green

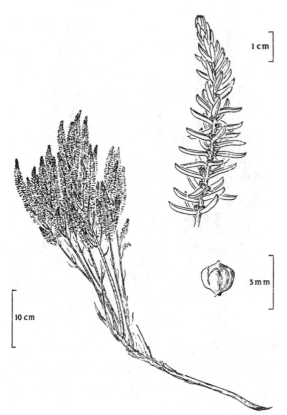

399. *Suaeda fruticosa* Forsk. 'Shrubby Seablite'   Green

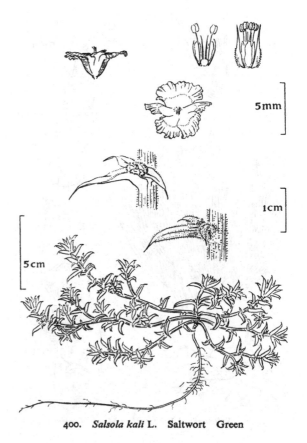

400. *Salsola kali* L.   Saltwort   Green

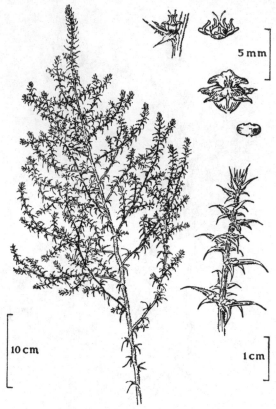

401. *Salsola pestifera* A. Nels.   Green

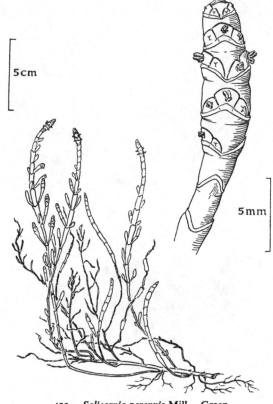

402. *Salicornia perennis* Mill.   Green

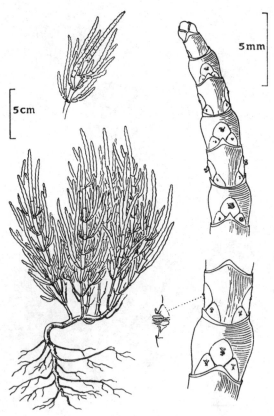

403. *Salicornia dolichostachya* Moss   Green

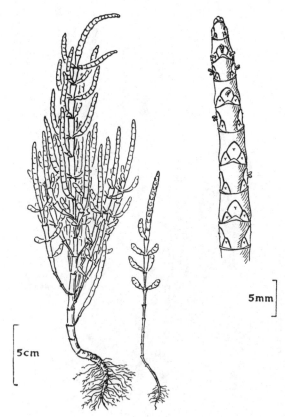

404. *Salicornia fragilis* P. W. Ball & Tutin   Green

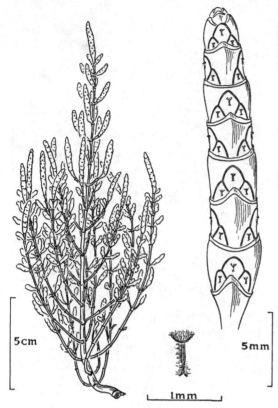

405. *Salicornia europaea* L.
Glasswort, Marsh Samphire   Green

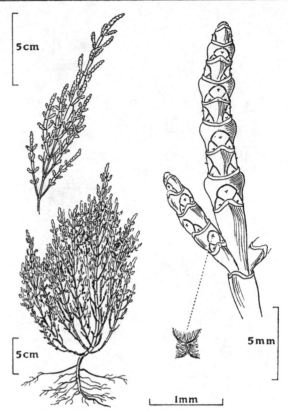

406. *Salicornia ramosissima* Woods   Reddish

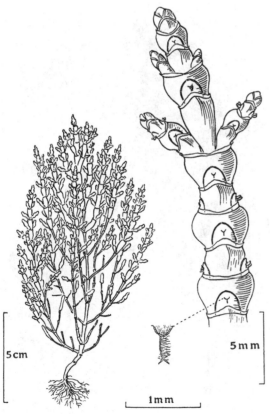

407. *Salicornia pusilla* Woods
(*S. disarticulata* Moss)   Greenish brown

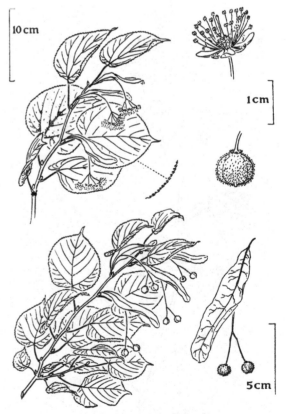

408. *Tilia platyphyllos* Scop.
'Large-leaved Lime'   Yellowish

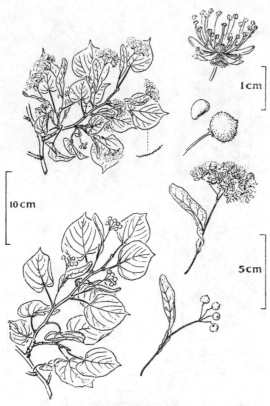

409. *Tilia cordata* Mill. 'Small-leaved Lime' Yellowish

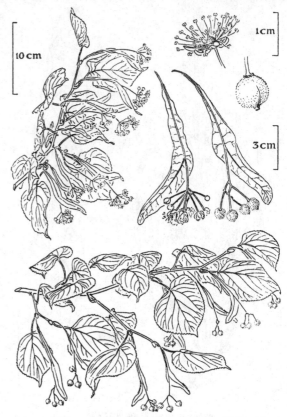

410. *Tilia × europaea* L. Common Lime Yellowish

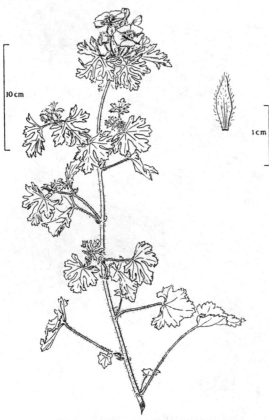

411. *Malva moschata* L. Musk Mallow Pink

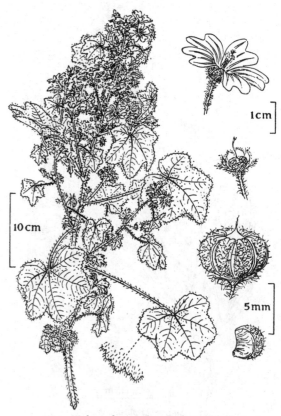

412. *Malva sylvestris* L. Common Mallow
Purplish pink

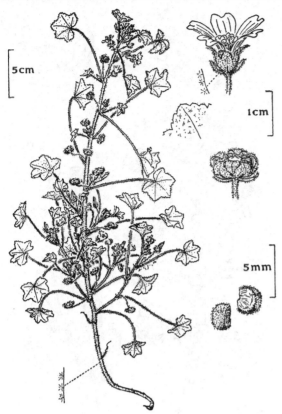

413. *Malva neglecta* Wallr. 'Dwarf Mallow' Pinkish

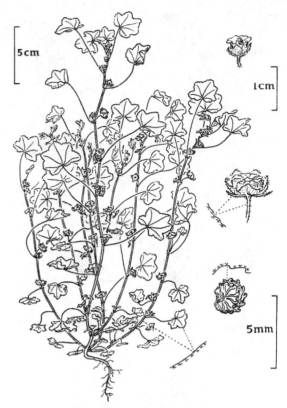

414. *Malva pusilla* Sm. Pinkish

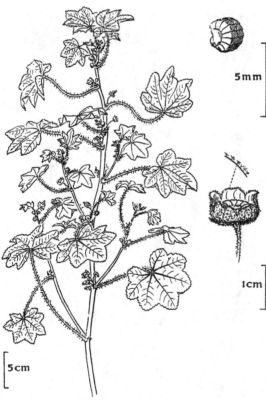

415. *Malva parviflora* L. Pinkish

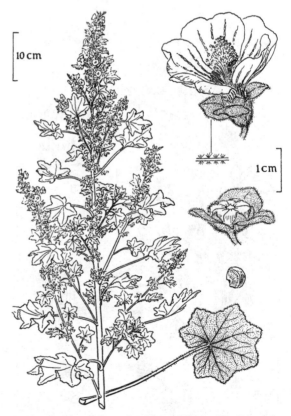

416. *Lavatera arborea* L. Tree Mallow Purplish pink

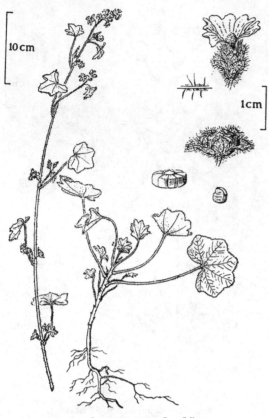

417. *Lavatera cretica* L.   Lilac

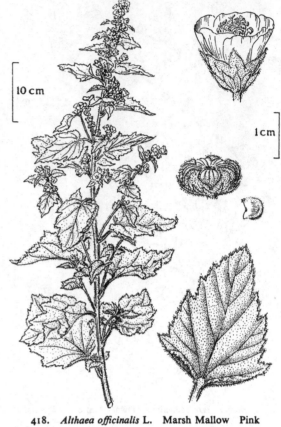

418. *Althaea officinalis* L.   Marsh Mallow   Pink

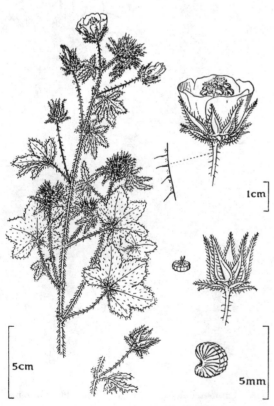

419. *Althaea hirsuta* L.   'Hispid Mallow'
         Pinkish purple

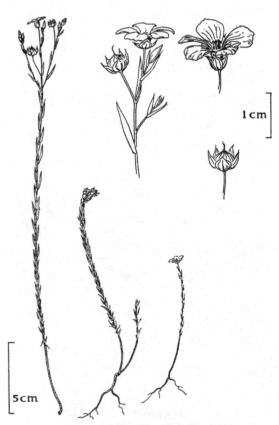

420. *Linum bienne* Mill.   'Pale Flax'   Blue

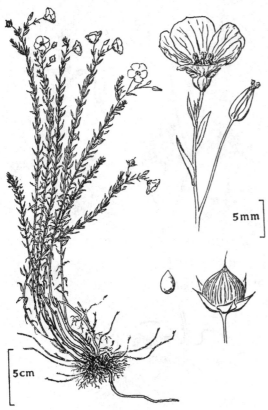

421. *Linum anglicum* Mill. 'Perennial Flax' Blue

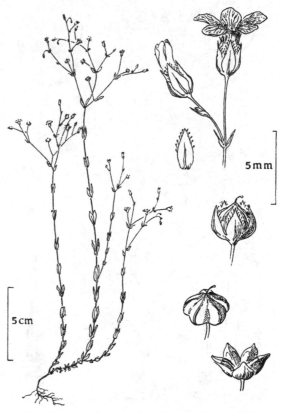

422. *Linum catharticum* L. 'Purging Flax' White

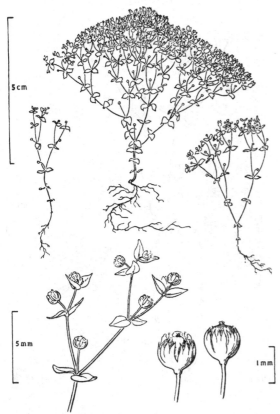

423. *Radiola linoides* Roth   All-seed   White

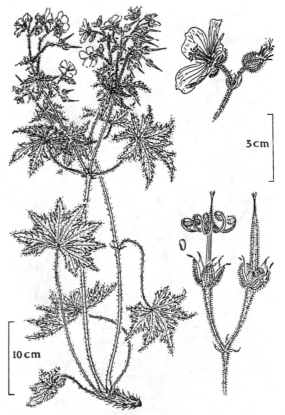

424. *Geranium pratense* L. 'Meadow Cranesbill'
Violet-blue

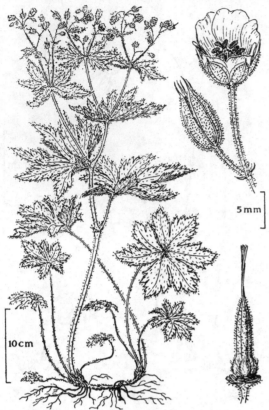

425. *Geranium sylvaticum* L. 'Wood Cranesbill'
Blue-violet

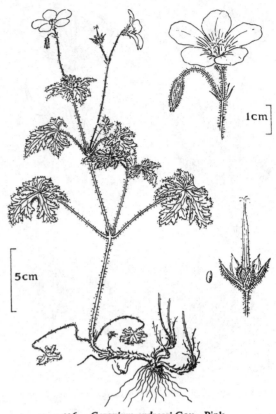

426. *Geranium endressi* Gay Pink

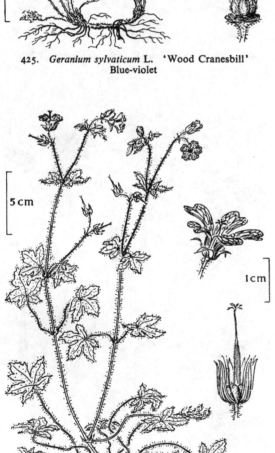

427. *Geranium versicolor* L. White or pale lilac

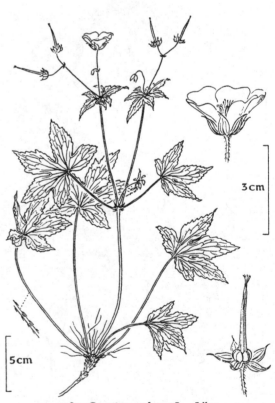

428. *Geranium nodosum* L. Lilac

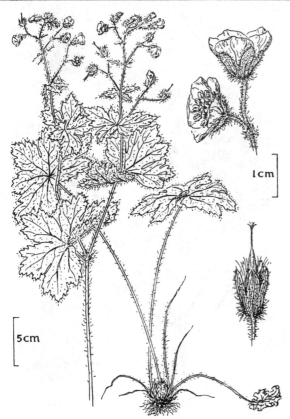

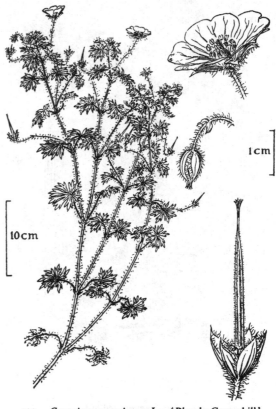

429.   *Geranium phaeum* L.   'Dusky Cranesbill'
Dark purple

430.   *Geranium sanguineum* L.   'Bloody Cranesbill'
Purplish crimson

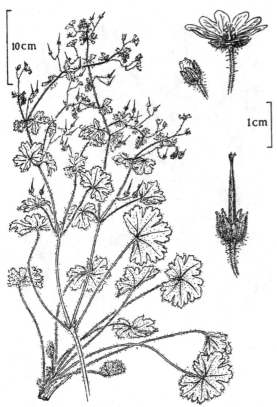

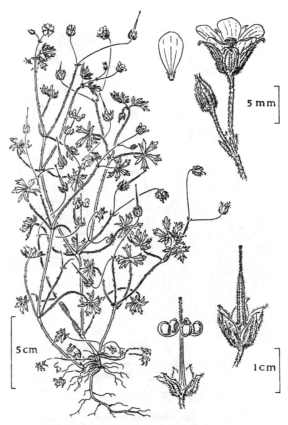

431.   *Geranium pyrenaicum* Burm. f.
'Mountain Cranesbill'   Purple

432.   *Geranium columbinum* L.
'Long-stalked Cranesbill'   Purplish pink

[109]

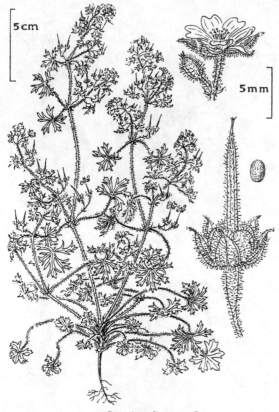

433. *Geranium dissectum* L.
'Cut-leaved Cranesbill' Reddish

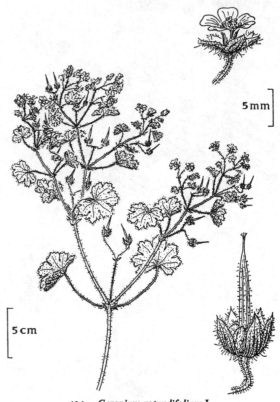

434. *Geranium rotundifolium* L.
'Round-leaved Cranesbill' Pink

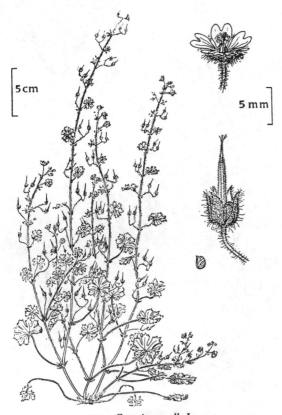

435. *Geranium molle* L.
'Dove's-foot Cranesbill' Purplish pink

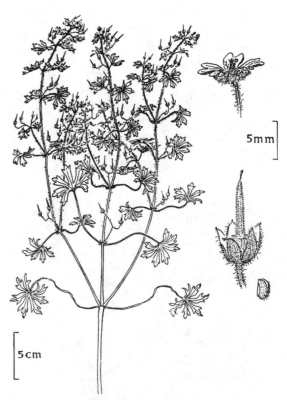

436. *Geranium pusillum* L.
'Small-flowered Cranesbill' Lilac

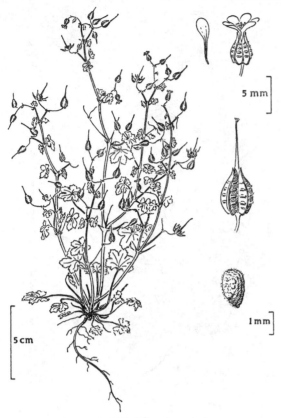

437.   *Geranium lucidum* L.   'Shining Cranesbill'   Pink

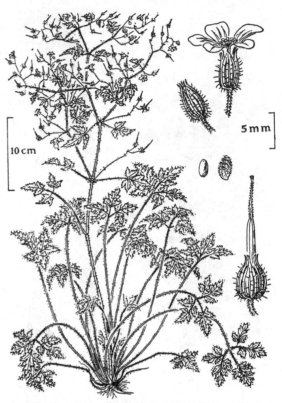

438.   *Geranium robertianum* L.   Herb Robert   Pink

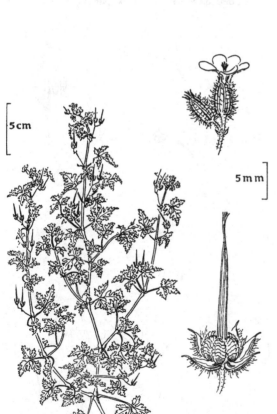

439.   *Geranium purpureum* Vill.   Pink

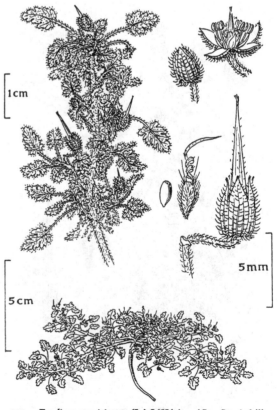

440.   *Erodium maritimum* (L.) L'Hérit.   'Sea Storksbill
Pink

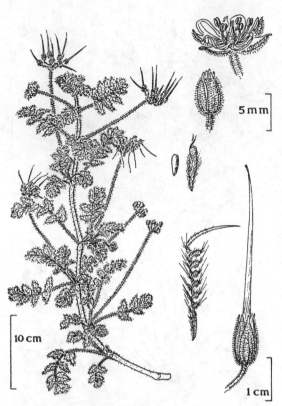

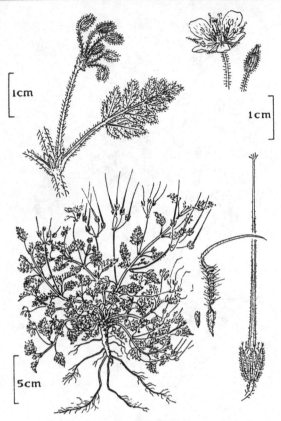

441. *Erodium moschatum* (L.) L'Hérit. 'Musk Storksbill'
Purplish pink

442. *Erodium cicutarium* (L.) L'Hérit. ssp. *cicutarium*
'Common Storksbill'   Pink, purplish or white

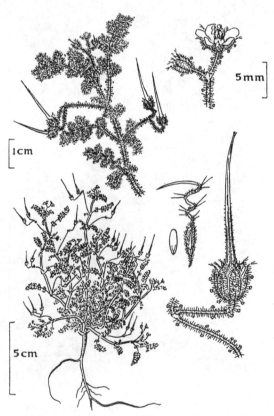

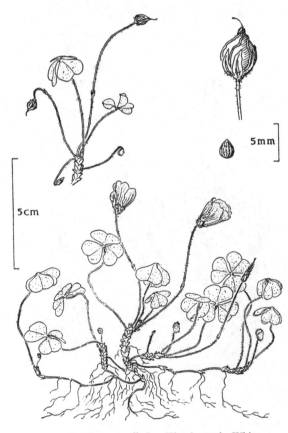

443. *Erodium glutinosum* Dumort.   Pink or white

444. *Oxalis acetosella* L.   Wood-sorrel   White

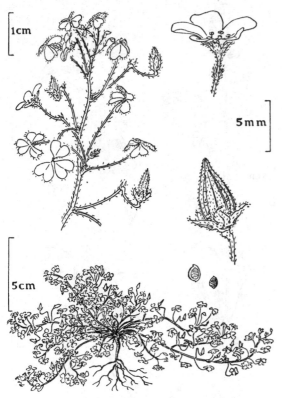

445. *Oxalis corniculata* L.
'Procumbent Yellow Sorrel'   Yellow

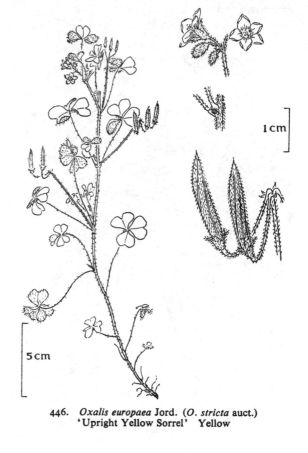

446. *Oxalis europaea* Jord. (*O. stricta* auct.)
'Upright Yellow Sorrel'   Yellow

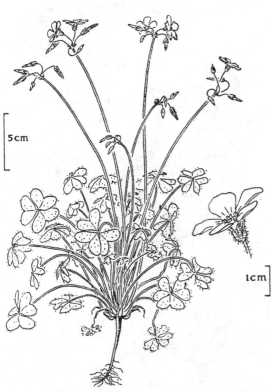

447. *Oxalis pes-caprae* L. (*O. cernua* Thunb.)   Yellow

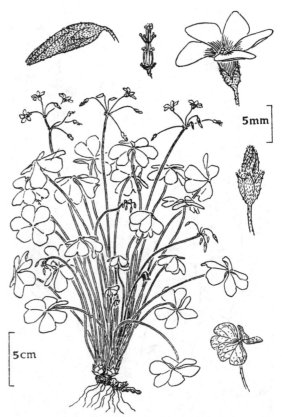

448. *Oxalis articulata* Savigny (*O. floribunda* Lehm.)
Purplish pink

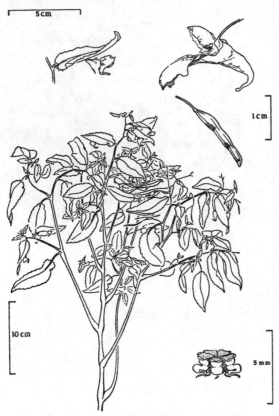

449. *Impatiens noli-tangere* L.   Touch-me-not
Yellow

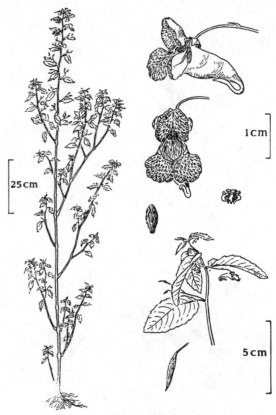

450. *Impatiens capensis* Meerb.   Orange Balsam
Orange

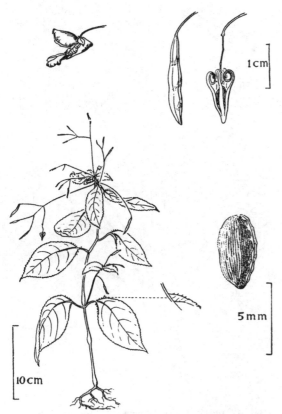

451. *Impatiens parviflora* DC.   'Small Balsam'
Pale yellow

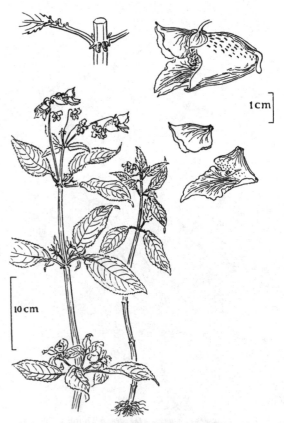

452. *Impatiens glandulifera* Royle
Policeman's Helmet   Purplish pink

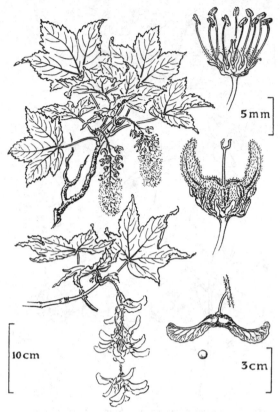

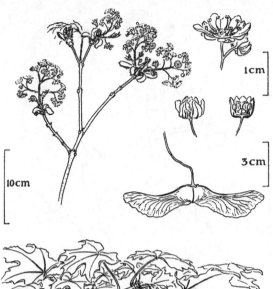

453. *Acer pseudoplatanus* L.   Sycamore
Yellowish green

454. *Acer platanoides* L.   Norway Maple
Greenish yellow

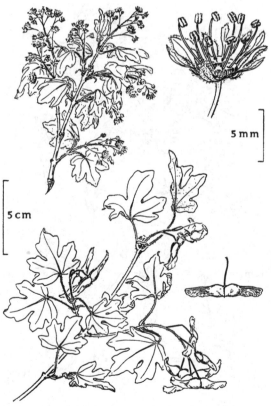

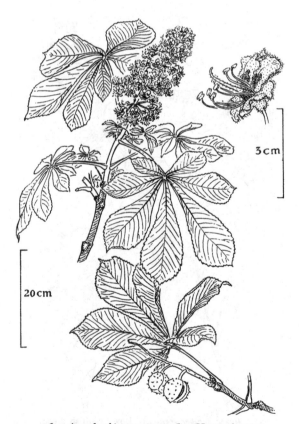

455. *Acer campestre* L.   Common Maple
Pale green

456. *Aesculus hippocastanum* L.   Horse-chestnut
White

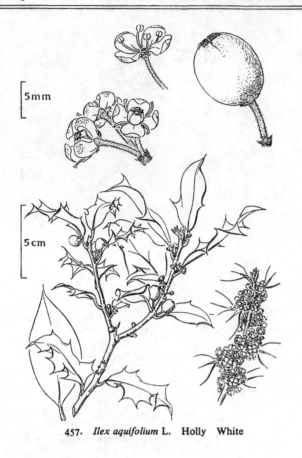

457. *Ilex aquifolium* L. Holly White

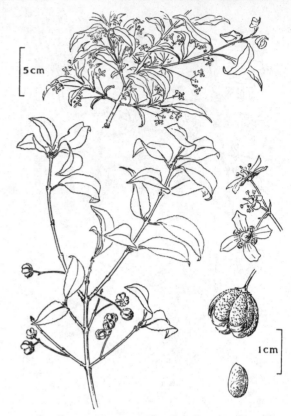

458. *Euonymus europaeus* L. Spindle-tree Greenish

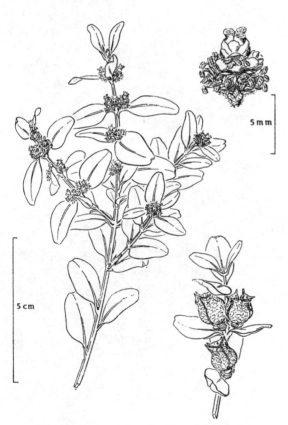

459. *Buxus sempervirens* L. Box Whitish green

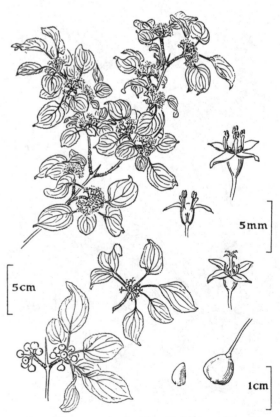

460. *Rhamnus catharticus* L. Buckthorn Greenish

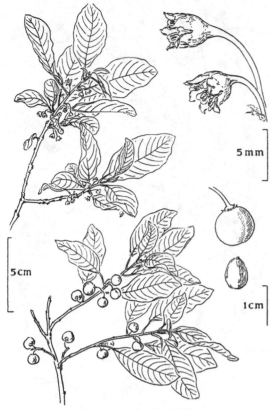

461. *Frangula alnus* Mill.
Alder Buckthorn, Black Dogwood   Greenish

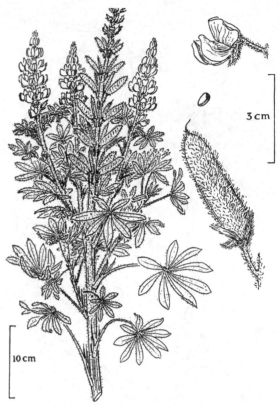

462. *Lupinus nootkatensis* Donn   Lupin   Blue

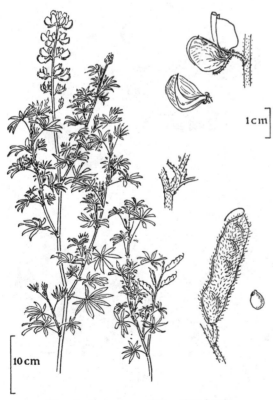

463. *Lupinus arboreus* Sims   Tree Lupin
Yellow or white

464. *Lupinus polyphyllus* Lindl.
Blue, pink or white

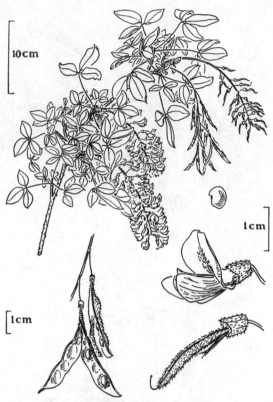

465. *Laburnum anagyroides* Medic.
Golden Rain, Laburnum   Yellow

466. *Genista tinctoria* L.   Dyer's Greenweed
Yellow

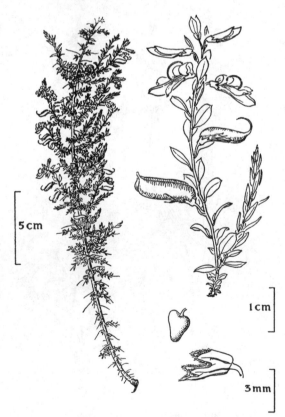

467. *Genista anglica* L.   Needle Furze, Petty Whin
Yellow

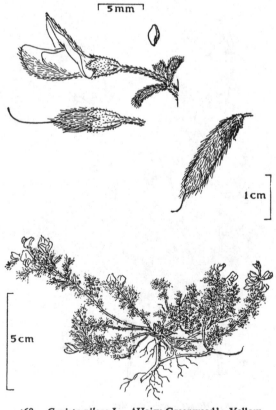

468. *Genista pilosa* L.   'Hairy Greenweed'   Yellow

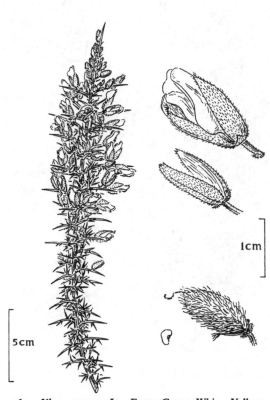

469.   *Ulex europaeus* L.   Furze, Gorse, Whin   Yellow

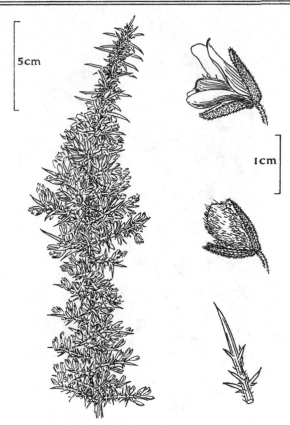

470.   *Ulex gallii* Planch.   'Dwarf Furze'   Yellow

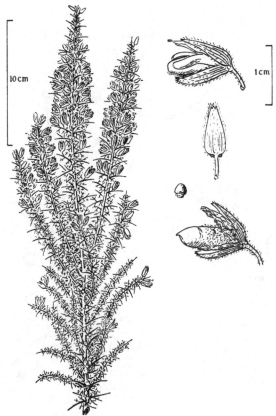

471.   *Ulex minor* Roth   'Dwarf Furze'   Yellow

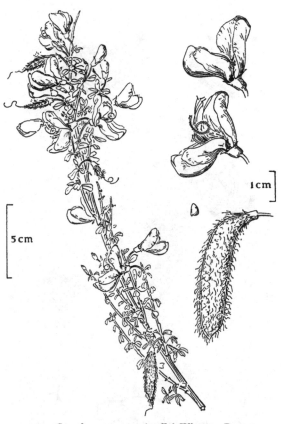

472.   *Sarothamnus scoparius* (L.) Wimm.   Broom
Yellow

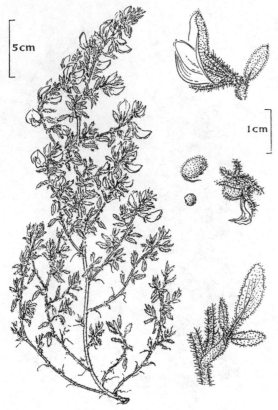

473. *Ononis repens* L. Restharrow Pink

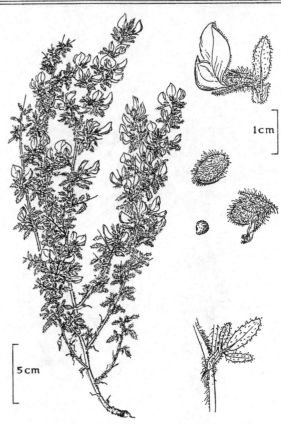

474. *Ononis spinosa* L. Restharrow Pink

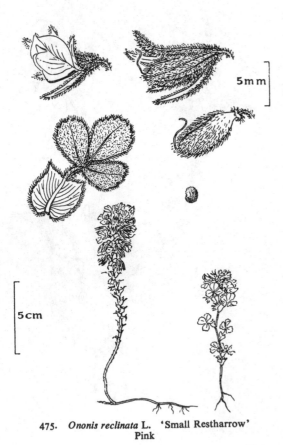

475. *Ononis reclinata* L. 'Small Restharrow'
Pink

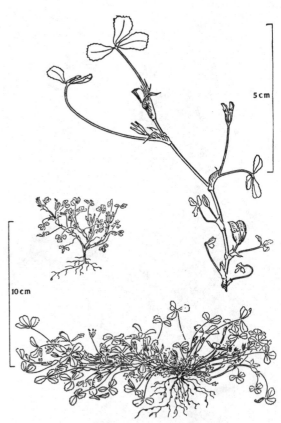

476. *Trigonella ornithopodioides* (L.) DC.
'Birdsfoot Fenugreek' Pinkish

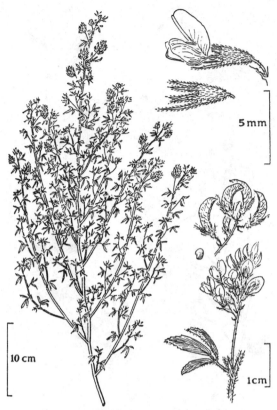

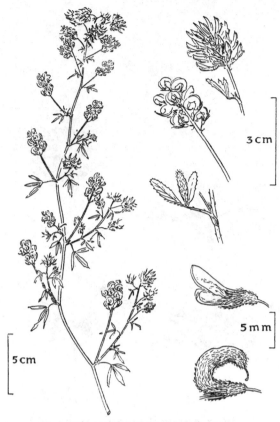

477.   *Medicago falcata* L.   'Sickle Medick'
Yellow

478.   *Medicago × varia* Martyn
Yellow, purple, dark green or black

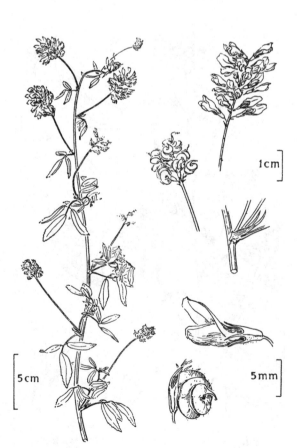

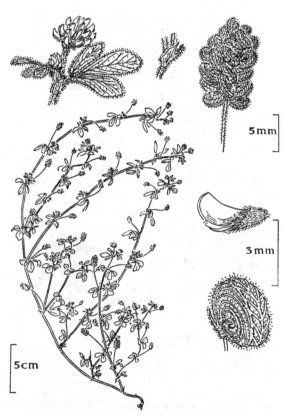

479.   *Medicago sativa* L.   Lucerne, Alfalfa   Purple

480.   *Medicago lupulina* L.   Black Medick   Yellow

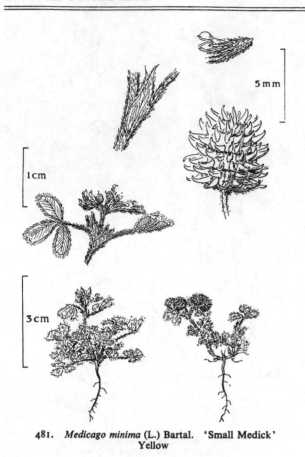

481.   *Medicago minima* (L.) Bartal.   'Small Medick'
Yellow

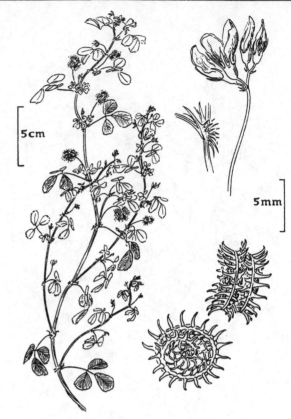

482.   *Medicago hispida* Gaertn.   'Hairy Medick'
Yellow

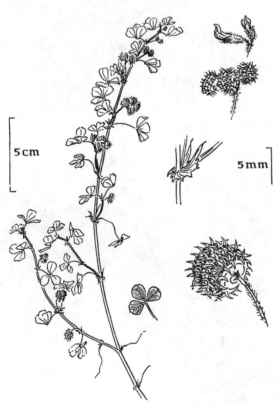

483.   *Medicago arabica* (L.) Huds.   'Spotted Medick'
Yellow

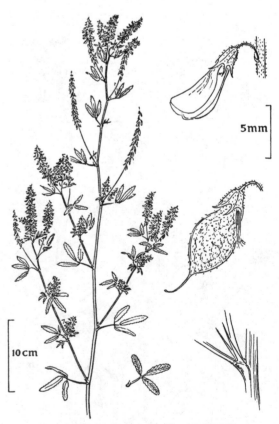

484.   *Melilotus altissima* Thuill.   'Tall Melilot'
Yellow

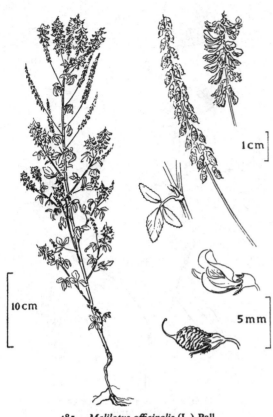

485. *Melilotus officinalis* (L.) Pall.
'Common Melilot'   Yellow

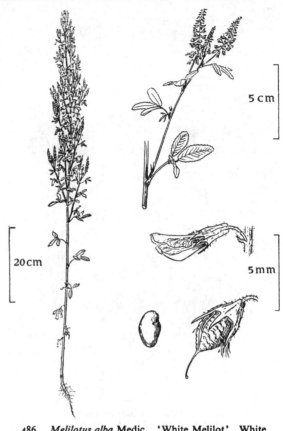

486. *Melilotus alba* Medic.   'White Melilot'   White

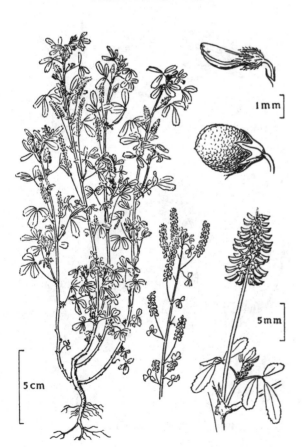

487. *Melilotus indica* (L.) All.
'Small-flowered Melilot'   Pale yellow

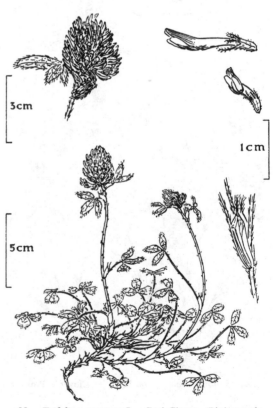

488. *Trifolium pratense* L.   Red Clover   Pink-purple

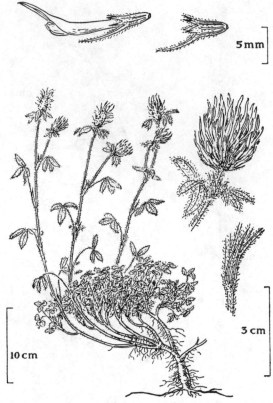

489.   *Trifolium ochroleucon* Huds.
'Sulphur Clover'   Whitish yellow

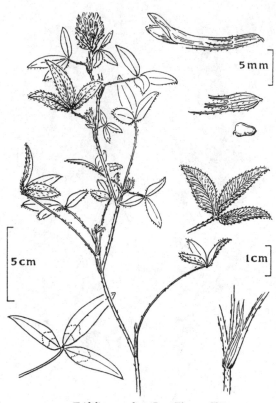

490.   *Trifolium medium* L.   Zigzag Clover
Reddish purple

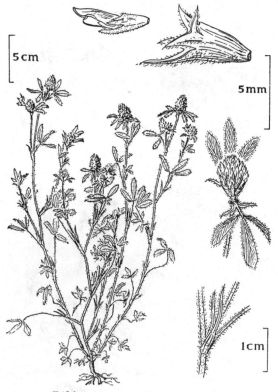

491.   *Trifolium squamosum* L.   'Sea Clover'   Pink

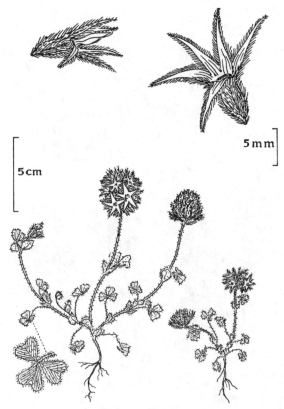

492.   *Trifolium stellatum* L.   Pink

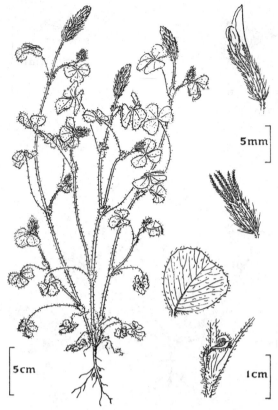

493.   *Trifolium incarnatum* L.   'Crimson Clover'
                 Crimson

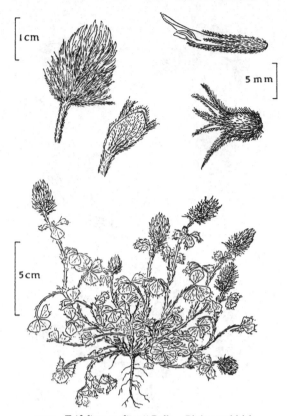

494.   *Trifolium molinerii* Balb.   Pink or whitish

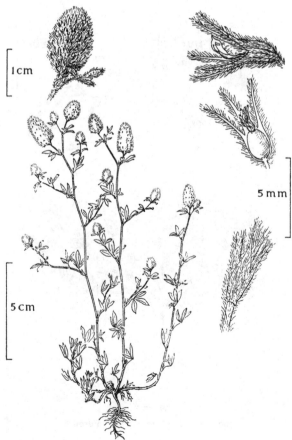

495.   *Trifolium arvense* L.   Hare's-foot   White or pink

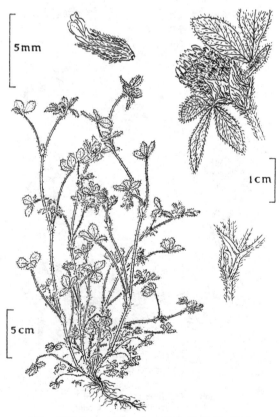

496.   *Trifolium striatum* L.   'Soft Trefoil'   Pink

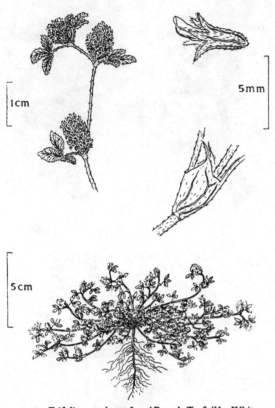

497.   *Trifolium scabrum* L.   'Rough Trefoil'   White

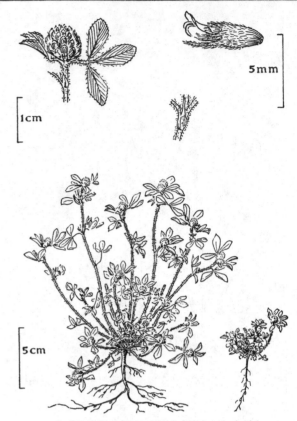

498.   *Trifolium bocconi* Savi   White or pinkish

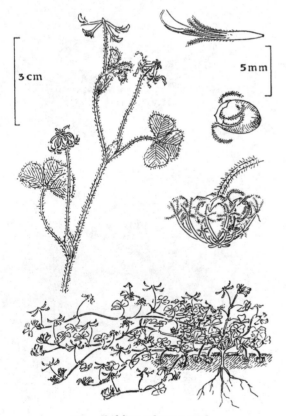

499   *Trifolium subterraneum* L.
Subterranean Trefoil'   Cream

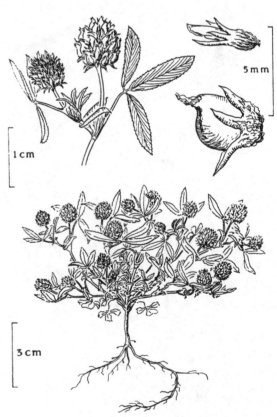

500.   *Trifolium strictum* L.   Purplish

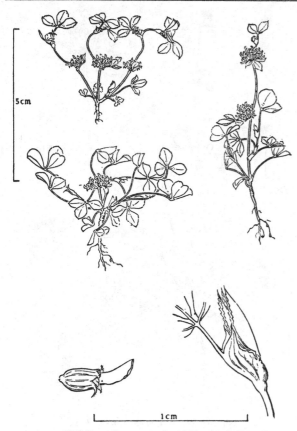

501.   *Trifolium glomeratum* L.   'Clustered Clover'
Purplish

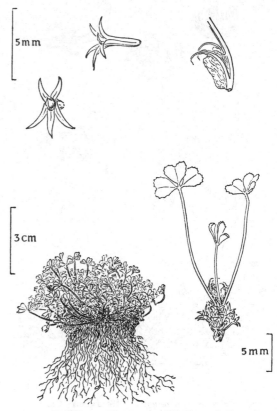

502.   *Trifolium suffocatum* L.   'Suffocated Clover'
Whitish

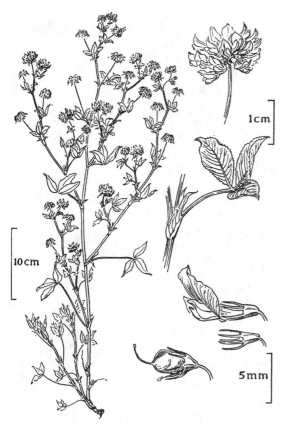

503.   *Trifolium hybridum* L.   Alsike Clover
White or pink

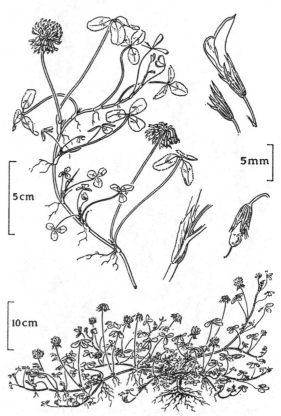

504.   *Trifolium repens* L.   White Clover, Dutch Clover
White or pink

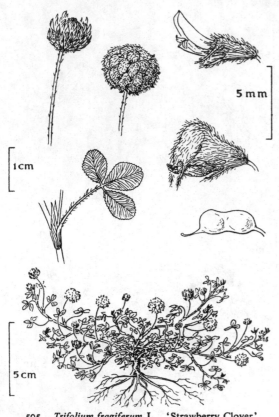

505. *Trifolium fragiferum* L. 'Strawberry Clover'
Pinkish or purplish

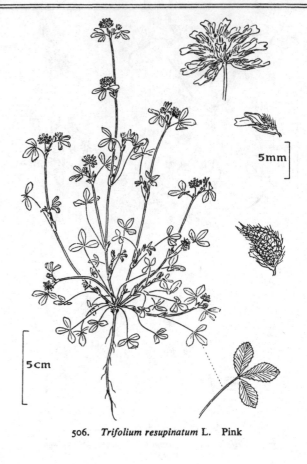

506. *Trifolium resupinatum* L.   Pink

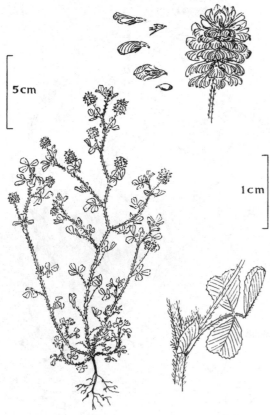

507. *Trifolium campestre* Schreb. 'Hop Trefoil'
Yellow

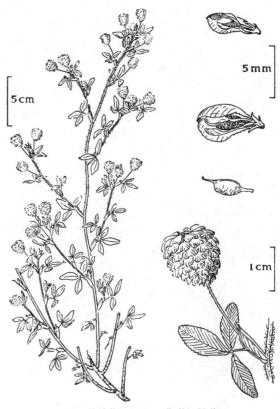

508. *Trifolium aureum* Poll.   Yellow

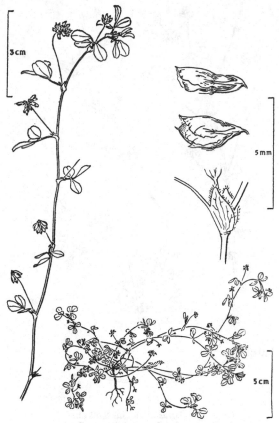

509. *Trifolium dubium* Sibth. 'Lesser Yellow Trefoil'
Pale yellow

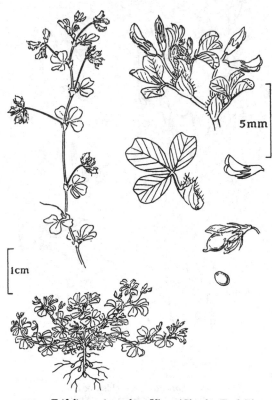

510. *Trifolium micranthum* Viv. 'Slender Trefoil'
Deep yellow

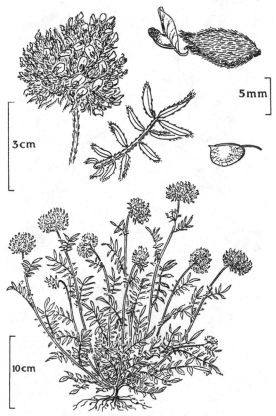

511. *Anthyllis vulneraria* L.
Kidney-vetch, Ladies' Fingers Yellow or red

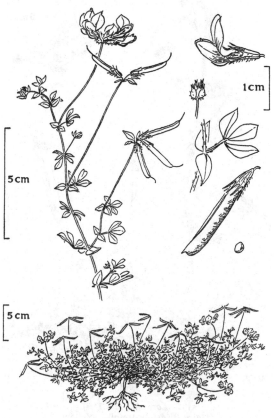

512. *Lotus corniculatus* L.
Birdsfoot-trefoil, Bacon and Eggs Yellow

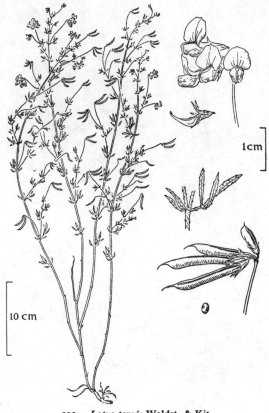

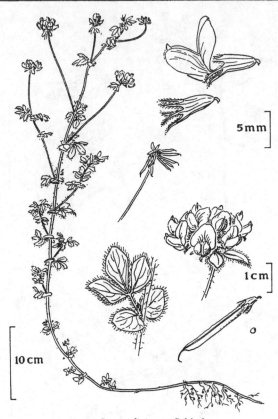

513.   *Lotus tenuis* Waldst. & Kit.
'Slender Birdsfoot-trefoil'   Yellow

514.   *Lotus uliginosus* Schkuhr
'Large Birdsfoot-trefoil'   Yellow

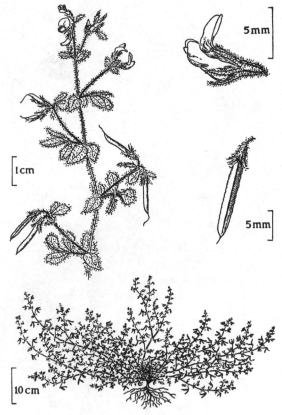

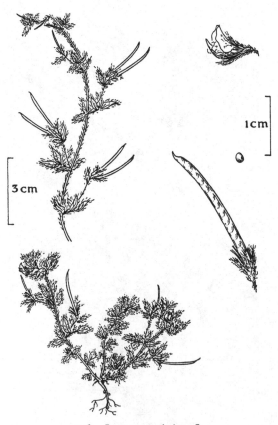

515.   *Lotus hispidus* Desf.
'Hairy Birdsfoot-trefoil'   Yellow

516.   *Lotus angustissimus* L.
'Slender Birdsfoot-trefoil'   Yellow

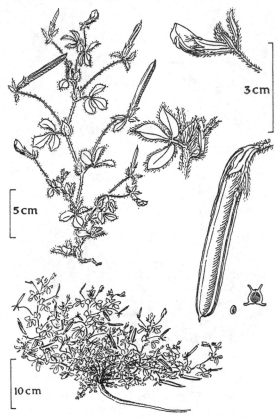

517.   *Tetragonolobus maritimus* (L.) Roth   Pale yellow

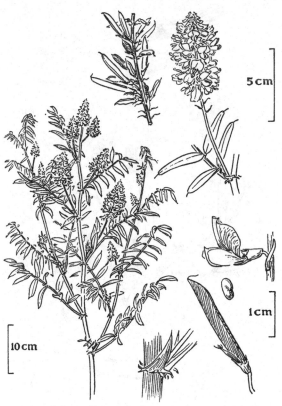

518.   *Galega officinalis* L.   Goat's Rue, French Lilac
White or lilac

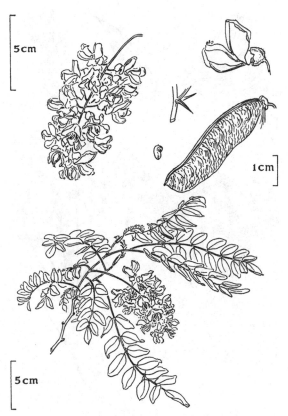

519.   *Robinia pseudoacacia* L.   Acacia   White

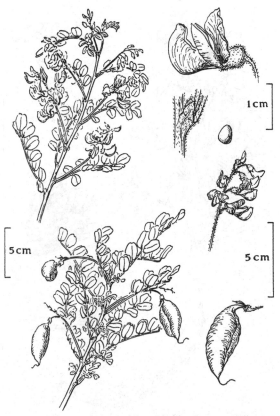

520.   *Colutea arborescens* L.   Bladder Senna   Yellow

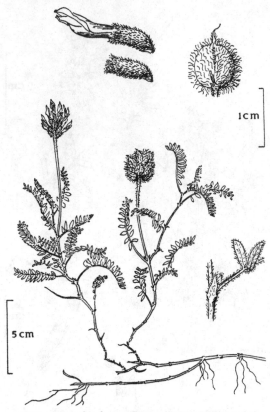

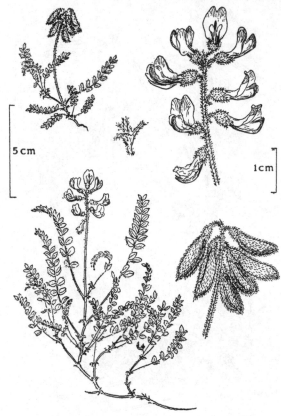

1cm

5cm

5cm

1cm

521.   *Astragalus danicus* Retz.   'Purple Milk-vetch'
Blue-purple

522.   *Astragalus alpinus* L.   'Alpine Milk-vetch'
Pale blue

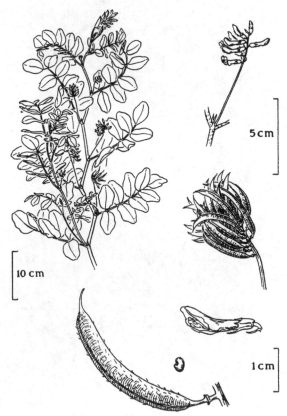

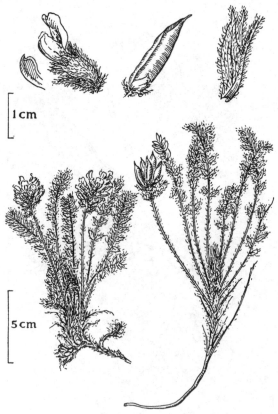

5cm

1cm

10 cm

5cm

1cm

523.   *Astragalus glycyphyllos* L.   Milk-vetch   Cream

524.   *Oxytropis halleri* Bunge   'Purple Oxytropis'
Purple

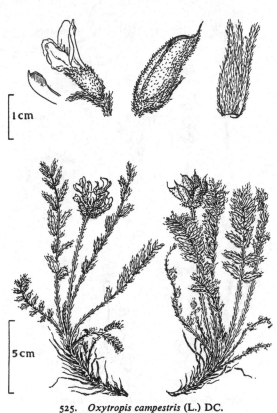

525. *Oxytropis campestris* (L.) DC.
'Yellow Oxytropis' Yellow

526. *Ornithopus perpusillus* L. Birdsfoot Pinkish

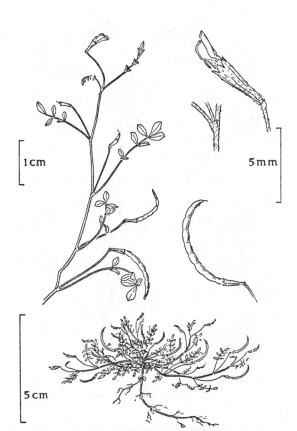

527. *Ornithopus pinnatus* (Mill.) Druce Yellow

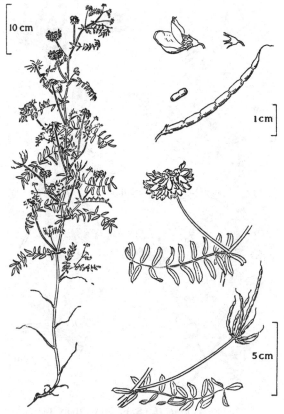

528. *Coronilla varia* L. Crown Vetch
White, purple or pink

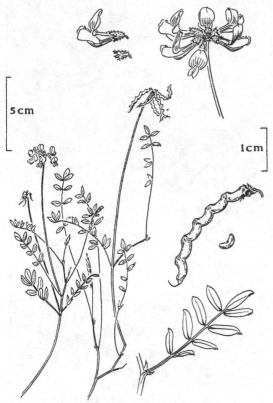

529. *Hippocrepis comosa* L. Horse-shoe Vetch
Yellow

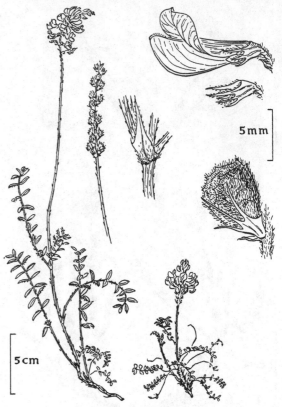

530. *Onobrychis viciifolia* Scop. Sainfoin
Red or pink

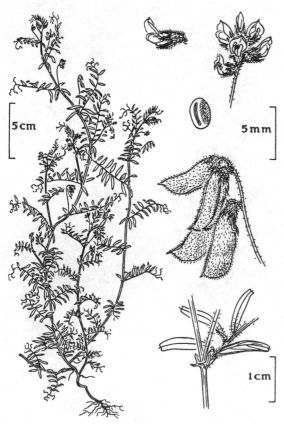

531. *Vicia hirsuta* (L.) S. F. Gray Hairy Tare
Whitish

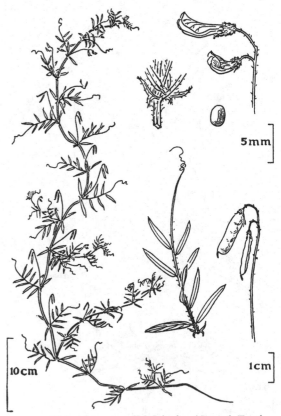

532. *Vicia tetrasperma* (L.) Schreb. 'Smooth Tare'
Pale blue

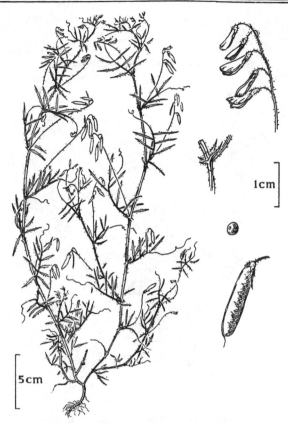

533.   *Vicia tenuissima* (Bieb.) Schinz & Thell.
         'Slender Tare'   Pale blue

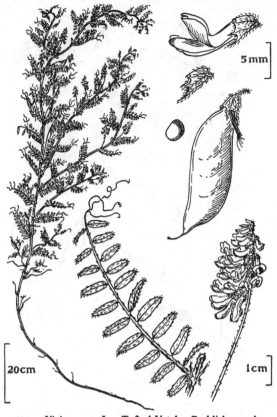

534.   *Vicia cracca* L.   Tufted Vetch   Reddish purple

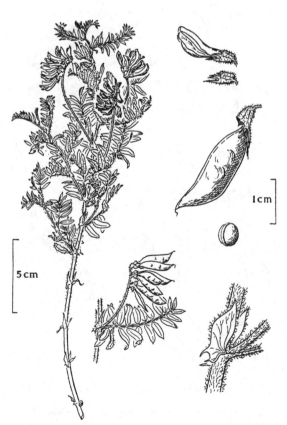

535.   *Vicia orobus* DC.   'Bitter Vetch'   Purplish white

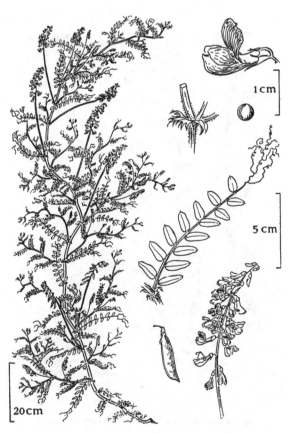

536.   *Vicia sylvatica* L.   'Wood Vetch'   Whitish

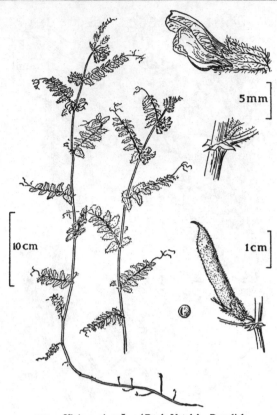

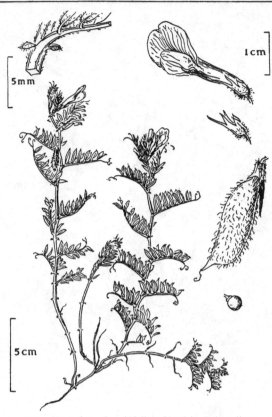

537.   *Vicia sepium* L.   'Bush Vetch'   Purplish

538.   *Vicia lutea* L.   'Yellow Vetch'   Pale yellow

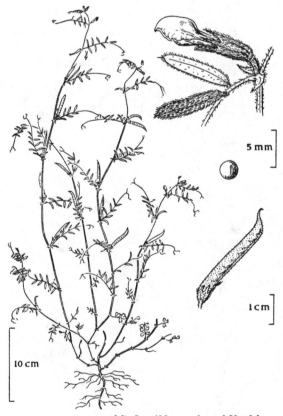

539.   *Vicia sativa* L.   'Common Vetch'   Purple

540.   *Vicia angustifolia* L.   'Narrow-leaved Vetch'
Purple

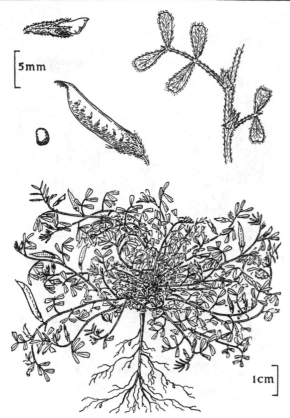

541. *Vicia lathyroides* L. 'Spring Vetch' Lilac

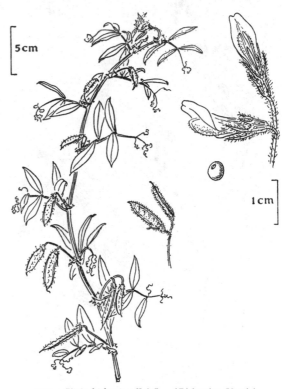

542. *Vicia bithynica* (L.) L. 'Bithynian Vetch'
Purple and white

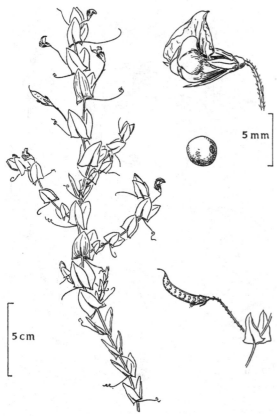

543. *Lathyrus aphaca* L. 'Yellow Vetchling' Yellow

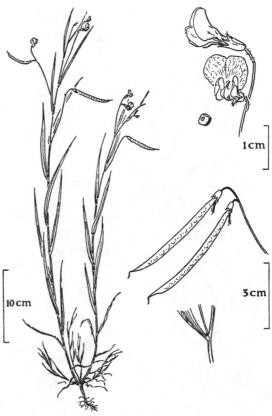

544. *Lathyrus nissolia* L. 'Grass Vetchling' Crimson

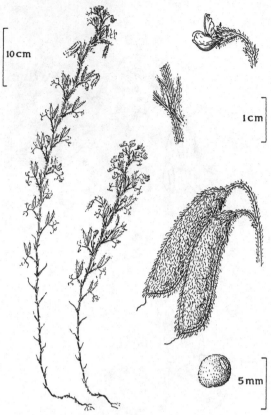

545. *Lathyrus hirsutus* L.   'Hairy Vetchling'
Crimson and blue

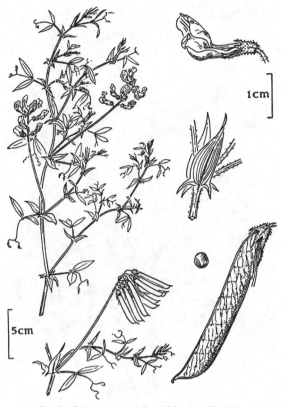

546. *Lathyrus pratensis* L.   'Meadow Vetchling'
Yellow

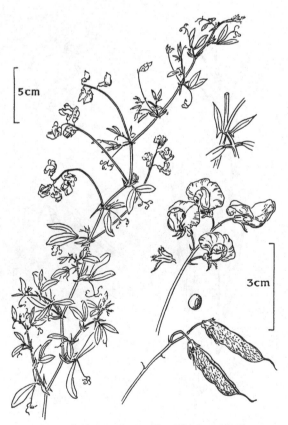

547. *Lathyrus tuberosus* L.   'Earth-nut Pea'
Crimson

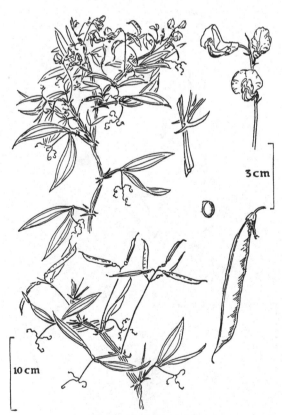

548. *Lathyrus sylvestris* L.
'Narrow-leaved Everlasting Pea'   Pink

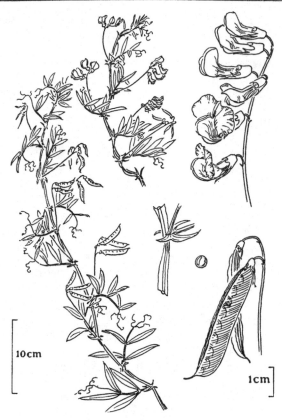

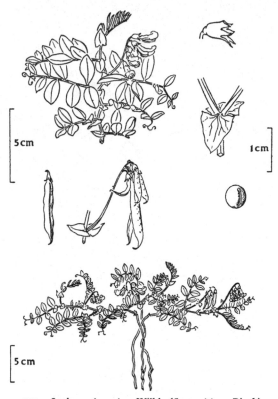

549.   *Lathyrus palustris* L.   'Marsh Pea'
Purplish blue

550.   *Lathyrus japonicus* Willd. (*L. maritimus* Bigel.)
'Sea Pea'   Purple to blue

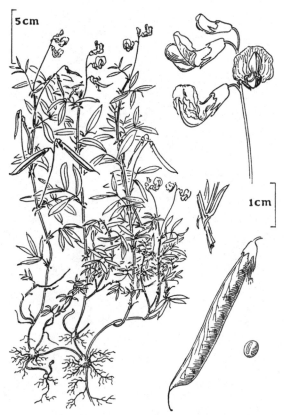

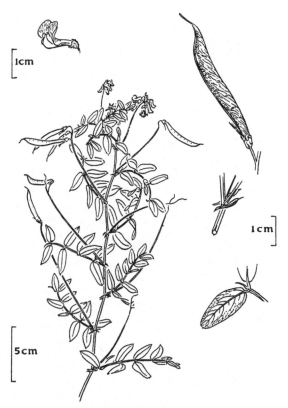

551.   *Lathyrus montanus* Bernh.   'Bitter Vetch'
Crimson

552.   *Lathyrus niger* (L.) Bernh.   Purple

INDEX